Ornithine Transcarbamylase

Ornithine Transcarbamylase

BASIC SCIENCE AND CLINICAL CONSIDERATIONS

PHILIP J. SNODGRASS, M.D.

Professor of Medicine Emeritus
Indiana University School of Medicine

KLUWER ACADEMIC PUBLISHERS
Boston/Dordrecht/London

Distributors for North, Central and South America:
Kluwer Academic Publishers
101 Philip Drive
Assinippi Park
Norwell, Massachusetts 02061 USA
Telephone (781) 871-6600
Fax (781) 681-9045
E-Mail: kluwer@wkap.com

Distributors for all other countries:
Kluwer Academic Publishers Group
Post Office Box 322
3300 AH Dordrecht, THE NETHERLANDS
Telephone 31 786 576 000
Fax 31 786 576 254
E-Mail: services@wkap.nl

Electronic Services < http://www.wkap.nl>

Library of Congress Cataloging-in-Publication Data

A C.I.P. Catalogue record for this book is available
from the Library of Congress.

Ornithine Transcarbamylase: Basic Science and Clinical Considerations by Philip J. Snodgrass
ISBN 1-4020-7683-5

Printed on acid-free paper.

Printed in the United States of America.

Dedicated to my wife, Marjorie,
for her patience and encouragement
and to my daughter Jennifer
for her editorial advice and guidance

Contents

Figures

Tables

Preface

Sir Hans Krebs described his discovery in 1932 of the urea/ornithine cycle in his autobiography, *Reminiscences and Reflections* (1981): "My idea was to use it (the tissue-slice technique) to study other metabolic processes, including synthetic ones, and as a first subject I chose the formation of urea in the liver. I was extraordinarily lucky in this choice because it led within twelve months to a major discovery—that of the ornithine cycle, the first 'metabolic cycle' to be identified"[(747)].

My own interest in this fascinating cycle began in 1964 when I chose ornithine transcarbamylase (OTC), the second enzyme in the cycle, to measure in human serum as a test for liver injury. Background reading led me to appreciate the major role that research on the arginine biosynthetic pathway in bacteria and fungi played in the development of molecular biology and molecular genetics. Part of this pathway evolved into the mammalian urea cycle in liver. Robert Schimke's publications in 1962–1964 showed that the rat liver urea cycle enzymes were induced by protein feedings. He demonstrated the role of both synthesis and degradation of the enzymes in the adaptation to the level of protein intake and to starvation, and also showed that glucocorticoids were necessary for maintaining the urea cycle activities. I decided that the problem of induction and suppression of the urea cycle in mammalian liver and the mechanisms by which the enzymes, their messenger-RNAs and genes were regulated had long-term possibilities as a research project for my laboratory. My colleagues and I pursued these problems from 1965 to 1993.

Clinicians, pediatricians and geneticists developed greater interest in the mammalian and human urea cycle because of acquired and genetic causes of hyperammonemia in humans. B. Levin and colleagues first reported OTC deficiency in 1962. Other researchers soon reported deficiencies of the other four urea cycle enzymes, of acetylglutamate synthetase and recently of transporters for ornithine/citrulline and for basic amino acids in liver, all of which caused hyperammonemia of varying severities. Our laboratory collaborated with many pediatricians and geneticists in their studies of urea cycle deficiencies. Our role was to assay the five cycle enzymes in liver, often in needle biopsies, and when possible to carry out kinetic studies on the human enzymes. As a gastroenterologist and hepatologist serving adult patients, I had daily experience with hyperammonemia due to acquired liver diseases but did not care for newborns or children with OTC deficiency.

I decided to focus this book on only one enzyme of the cycle, OTC, because I had done more research on it than I had on the other enzymes and genetic OTC deficiency appeared to be the most common defect in the urea cycle. By compiling in one volume the widely scattered information on OTC, both clinical and basic, covering 1962–2002, I hope to help scientists and clinicians find a ready source of useful information about this interesting and important enzyme.

Acknowledgments

The author acknowledges the contributions of my colleagues, Renee C. Lin, PhD, and Corinne Ulbright, PhD; of my dedicated research fellows; of the many excellent technicians in our laboratory over a 30-year period; of many collaborators in the U.S. from departments of pediatrics, genetics and medicine; and of Patricia Lund, DPhil, of the Metabolic Research Laboratory in Oxford University. The author wishes to thank Thomas D. Hurley, PhD, Professor, Department of Biochemistry and Molecular Biology, Indiana University School of Medicine, for his collaboration in preparing Chapter 4 and reviewing Chapters 1–6. The author appreciates the critical reading of Chapters 1–6 by Robert A. Harris, PhD, Showalter Professor and Chairman of the Department of Biochemistry and Molecular Biology, and of Chapters 7 and 8 by Rebecca S. Wappner, MD, Professor of Pediatrics and Director of Metabolism/Genetics, JW Riley Children's Hospital, Indiana University School of Medicine. Any omissions or misinterpretations throughout the text rest entirely with the author.

1

Gene Structure, Regulation and Function

The human OTC gene has been localized to the short arm of the X chromosome at Xp21.1[1], approximately 12 centimorgans (cM) closer to the centromere than the gene for Duchenne muscular dystrophy. The gene was located in normal cell lines by *in situ* hybridization to mitotic chromosome spreads, employing a nearly full-length cDNA probe. Finer localization was accomplished by using the probe on chromosomes with known deletions in the p21 region[1]. Because it is located on the X chromosome, lyonization of the OTC gene in females due to random inactivation of one of the X-chromosomes[2] leads to a mosaic pattern in the liver which was elegantly demonstrated in the liver of a female heterozygous for OTC deficiency[3].

In the liver, OTC mRNA constitutes only 0.2 percent of all polyA-mRNAs, making cloning of this mRNA difficult. Cloning of a rat liver cDNA was finally accomplished by Horwich et al.[4] using immuno-affinity-enriched polysomal OTC mRNA[5]. A rat liver cDNA was produced (pOTC-1) which included codons of 52 C-terminal amino acids, 355 nucleotides of the 3′-region and a second unidentified 400 base pair (bp) cDNA upstream from the OTC. Part of this cDNA containing 221 bp of C-terminal coding sequence was used as a probe for isolating a full-length human OTC cDNA which was sequenced by Horwich et al.[6] Davies also isolated partial rat and human OTC clones[7]. Hata isolated and sequenced complementary and genomic DNA clones for human OTC[8]. The gene spans a region of 73 kilobases (kb) and contains 10 exons interrupted by 9 introns varying from 80 base pairs to 21.7 kb in size. A schematic representation of the OTC linear gene structure is shown in Fig. 1-1[9]. Hata et al.[8] found 20 nucleotide substitutions in their sequence compared with that of Horwich[6], 10 in the coding region and 10 in the 3′-untranslated region. Only 6 nucleotide differences related to amino acid changes in the cDNA. Hata found codon 14 in the cDNA for the OTC monomer could be AGA (**R**) whereas Horwich found that the codon was AAA (**K**). The five other nucleotide and amino acid differences were (Hata vs. Horwich): (TTA)**L69F**(TTT),(CTT)**L79P**(CCT), (TGG)**W161C**(TGT), (ATC)**I162F**(TTC)and (CAA)**Q238R**(CGA). All 6 amino acid differences were considered to be polymorphisms. Hata also sequenced the 5′-flanking region of the human OTC gene[10]. A single transcriptional start site was not determined. There are two pairs of candidates for CAAT and TATA boxes.

The cDNA for the human pOTC gene contains a 96 base-pair region which codes for a 32 amino acid leader peptide and a 966 base-pair (322 amino acid monomer) sequence[6]. Asparagine is the N-terminal amino acid of the mature monomer and phenylalanine the C-terminal amino acid. The leader peptide has a uniform distribution of arginines at positions −7, −10, −18, −27, a histidine at −15, and a cysteine at −6. There are no acidic residues. No membrane spanning

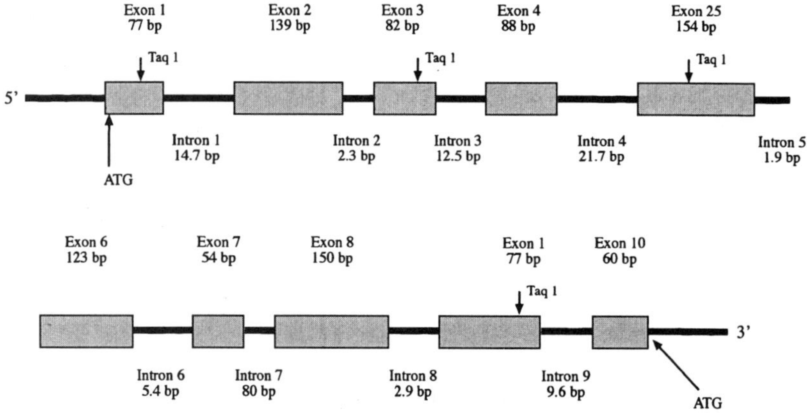

Figure 1-1. Schematic illustration of the structure of the OTC gene[9]. (Reprinted by permission.)

hydrophobic region was isolated so it does not resemble the signal sequence of secreted proteins. Two highly conserved regions among the human, rat and mouse genes in the monomer sequence are positions 54 to 60 and 268 to 273 which are thought to be the sites of binding of carbamyl phosphate and ornithine respectively. OTC cDNAs from the rat[4,11,12] and the mouse[13] have been isolated and sequenced. Sequencing of the rat OTC shows 93% homology with the human subunit but only 69% homology in the leader peptide[12,14]. In the mouse gene the transcription start site is 136 bp upstream from the translation initiation codon. An 800-bp fragment of the 5′ flanking region allows the gene to be transcribed in hepG-2 cells but not in 3T3 fibroblasts, showing that hepG-2 cells contain a liver cell specific regulator sequence[15]. The OTC mRNAs for the human, rat and mouse all contain at least 1572 nucleotides including 150 bases of 5′ untranslated sequence, 1062 bases of coding sequence and 360 bases of 3′-untranslated sequence.

Using the human 5′-promoter sequence of 589 bp reported by Hata[10], which includes exon-l, we have done computer matches for the consensus glucocorticoid receptor binding (GRE) sequence (TGTTCT) and no homologies were found. However, Takiguchi et al.[16] found one GRE on the sense strand of intron 1 of rat OTC (+499 to +514) and another on the antisense strand of intron-l (+290 to +275). We also searched for the consensus cyclic AMP response element (CRE) sequence (TGACGTCA). No good matches were found for this promoter sequence. However AP2, whose consensus sequence, CCCCAGGC, is also a cyclic AMP-responsive transcriptional factor, showed a 7 out of 7 nucleotide match over sequence −585 to −579. Antisense GRE, CRE, and AP2 sequences were not found. A 15 base-pair urea cycle enhancer sequence GAGCAGCCTGCCCTG was reported by Ohtake[17] in the 5′-region of the rat arginase gene, and a closely matched sequence called the urea cycle element (UCE) was found in the 5′-region of OTC and argininosuccinate synthetase. The core element of TGCCCT may account for coordinate induction of the ornithine cycle and related enzymes like ornithine aminotransferase by dietary protein[18].

DNAse-I footprint analysis of the rat promoter shows four protected regions[19]. Region A overlaps a previously identified negative element and regions B and C overlap the previously identified positive elements. The factors on B and C appear to be related to the chicken ovalbumim upstream promoter (COUP) transcription factor that is a member of the steroid super family. DNAse I foot-prints of the enhancer also showed four protected regions. Regions one and four are the same when brain or liver extracts are used but regions two and three are protected only with a liver extract. Gel shift competition showed that the regions 2 and 3 bind a C/EBP transcription protein. Further studies[20] showed that two sites of the promoter region and two sites on the 11-kb upstream enhancer were recognized by both the COUP transcription factor and hepatocyte nuclear factor 4 (HNF-4). HNF-4 activates the gene in the liver while COUP-TF repressed expression of the promoter[20].

Murakami et al.[21] studied the promoter region and the enhancer region out to −13.8 kb on the 5′-end. They found that the 1.3 kb promoter is active only in a liver cell or liver-derived tumor and not in a fibroblast or non-liver tumor cell showing that there is some liver specific controlling sequence in the promoter region. Murakami found that there was a negative acting cis element −222 to −162 bp above the initiation site and two positive cis acting elements at −112 to −85 and −42 to −15 bp. The enhancer region from −13.8 to −1.6 kb proved to have a 2.7 kb region between −13.8 and −11.1 kb which when transfected into hepG2 cells showed a 10-fold increase in transcription in the positive and an 18-fold increase in transcription in the negative direction. Between −11.1 and −7.9 kb the sequence only increased transcription 3.2 times. This enhancer does not function in Chinese hamster ovary (CHO) cells so the enhancer is liver specific. In a 230 base-pair sequence just above −11.1 kb two liver specific sites have been found related to a C/EBP transcription factor. Nishiyori[22] showed that in a 110-base pair region of the enhancer, two HNF-4 and two C/EBPβ sites account for liver-selective transcription enhancement.

In the *ARG3* (OTC) gene of *S. cerevisiae* two arginine boxes (ABI and AB2) have been identified immediately downstream from the TATA box. These promoter sites are thought to be the binding sites for the arginine repressor proteins (*ARGRI* and *ARGRII*)[24]. We cannot identify AB1 or AB2 sequences in the human OTC 5′ region[10]. However, a 149bp arginine repressor sequence for argininosuccinate synthetase has been identified in human RPMI 2650 cells in the promoter region[25]. It is not known what part of the 149bp is essential for arginine repression. We have not found any part of this 149bp sequence in the human OTC promoter[10] even though an arginine deficient diet induces (derepresses) the first four enzymes of the urea cycle while suppressing arginase[26,27] and arginine deficient media derepress non-liver cell lines[28].

Tissue and sex specific regulation of the OTC gene is accomplished by methylation of G + C-rich islands 12 kb 5′ of the OTC-coding region in mouse DNA. This MspI site was hypomethylated in male liver DNA but hypermethylated in male kidney DNA where OTC is not expressed[29]. The same site is almost fully methylated on the inactive X chromosome of female mouse liver.

Besides those OTC gene sequences listed in references 4 to 24, we have collected all of the other OTC gene sequences reported in the literature from 1983–2002 after a search of databases in PubMed indexed for MEDLINE and listed them in the references to this chapter in order by date of publication[30–51,708]. We will point out a number of important findings from a comparison of these sequences. Fungal OTCs contain no introns because fungi cannot splice exons while deleting introns. The ascomycetous fungus *S.cerevisiae* OTC has no leader peptide[23] and therefore is cytoplasmic whereas its close relative *A.nidulans*[32] OTC has a leader peptide and is able to enter the mitochondria as in ureotelic animals. The fungi which have leader peptides that must be cleaved to enter mitochondria usually contain a putative signal-peptide cleavage signal which occurs 1–3 amino acids upstream of the ATG (M) beginning of the OTC monomer. Examples are RSYSS in *A.nidulans*[32], RLYSS in *A.niger*[33], RFFSS in *P.tannophilus*[38] and RQYSS in *A.terreus*[46]. These cleavage motifs are not found in mouse, rat or human leader sequences.

We have compared the nucleotide sequences of 23 species in the carbamyl phosphate (CP) binding site (S58-T59-R60-T61-R62) and the ornithine binding site (H270-C271-L272-P273) using the numbering of the human OTC monomer[6]. In 23 species where these nucleotide sequences are available, HCLP is invariantly coded for by the various triplet codons for these four amino acids. In OTCs from 4 mammalian livers, from a chicken kidney, from 6 yeast/fungi and from 10 bacteria the various triplets always coded for the amino acids STRTR at the CP binding site. Only two exceptions occurred, one in the plant *Arabidopsis thaliana* where the sequence is SMRTR[50]. The other exception was the OTC of *Pseudomonas syringae* pv. *phaseolicola* where the nucleotides coded for SGRTS[39]. The other, housekeeping OTC gene in this bacterium codes for the STRTR motif. Phaseolotoxin is a tripeptide produced by this bacterium which produces a halo blight disease of beans. The toxin binds to the STRTR binding site of CP, which makes it cytotoxic to plants and bacteria that use the STRTR binding motif. It is postulated that this resistant *Ps. syringae* is protected against its own toxin because the toxin binds weakly to the SGRTS motif, but CP still binds well enough to give OTC activity and arginine synthesis(39,40).

Based on the amino acid sequences available in 1995, an evolutionary tree of the OTC protein family constructed by Ruepp et al.(43) showed that an archaeabacterium like catabolic *H. salinarium* was more closely related to vertebrates than it was to prokaryotes like *E.coli* and catabolic *P.aeruginosa.* Also anabolic *P.aeruginosa* was descended from the same precursor as *H.salinarium* and *B.subtilis.* The *argF* gene of anabolic *P.aeruginosa* differed from the *arcB* gene of catabolic *P.aeruginosa* which was more similar to *argF* and *argI* genes of *E.coli*(35). Two plant OTCs were closer evolutionarily to prokaryotes than to fungi and vertebrates(50). More extensive comparisons of amino acid sequences in OTCs of 33 species and their evolution will be discussed in Chapter 4.

Commentary. The OTC gene is located on the X-chromosome, is large and complex, and has maintained amazingly similar sequences in the mature OTC subunit across the evolutionary scale, especially in the carbamyl phosphate and ornithine binding sites. The human 32-amino acid leader peptide contains arginines at critical positions. Control elements in the 5′-region include liver specific elements, a urea cycle element found in other waste nitrogen genes, and glucocorticoid receptor elements.

2

Synthesis, Processing and Assembly

OTC protein is synthesized on free polysomes from its mRNA in the liver cytosol[52] resulting in a 32 amino acid leader peptide and a 322 amino acid subunit called pOTC[6,53,54]. As the nascent peptide of pOTC is released from the ribosome, it associates with several different molecular chaperones which bind the precursor and prevent its aggregation or irreversible folding. These include a heterodimeric protein termed mitochondrial import stimulating factor (MSF)[55,56], a constitutive heat shock cognate 70 protein (hsc70)[57], a human homolog of bacterial heat shock protein DnaJ called HSDJ which is farnesylated to become active[58] and a presequence binding factor (PBF)[59]. This pOTC-chaperone complex acts as a transport particle and moves to the mitochondria, possibly along microtubules. MSF cleaves ATP and delivers pOTC to a group of Tom subunits (transport across the outer membrane) at translocation contact sites where the outer and inner mitochondrial membranes are in apposition[56,60]. Polyamines may play a role in binding of pOTC to the outer membrane[61,62]. The import receptor for pOTC-MSF is a Tom70-Tom37 complex and is transferred to Tom22-Tom20 subunits where it enters a protein channel made up of Tom subunits 40,38,7,6 and 5[56,63]. A Tom34 may be a novel translocase in human mitochondria[64]. After pOTC crosses the outer membrane it is pulled across the inner membrane by the electrochemical potential, probably moving through another channel composed of subunits of the Tim complex (transport across the inner membrane)[56]. Tim44 binds to the chaperones mitochondrial hsp70 (mhsp70) and to mGrpE. ATP hydrolysis furnishes the energy to pull the pOTC across the inner membrane[56]. Maximal transport and processing of pOTC requires 120 mM K^+ and 0.8–1.6 mM Mg^{++}, probably as co-factors for the ATPase[65,66]. Once in the matrix pOTC is kept in an unfolded state by mhsp60[67] and chaperonin 10 (cpn10)[68]. In the mitochondrial matrix, pOTC first encounters a neutral protease[69] called the mitochondrial processing peptidase (MPP) that cleaves the first 24 N-terminal amino acids. Numbering from the N-terminal methionine the cleavage site is between asn 24 and phe 25, one bond away from a critical arg 23[70]. This results in a 37 kDa intermediate form[52,54] (iOTC). The intermediate form is not produced unless pOTC penetrates the outer and inner mitochondrial membranes by a process that requires energy generated by oxidative phosphorylation[65,71]. Normal inner membrane charge must be present, negative on the inner side and positive on the outer side of the membrane, for the intermediate OTC to be generated[71]. Intermediate OTC then encounters a second matrix protease that is zinc dependent[72]. This mitochondrial intermediate peptidase (MIP) removes the next 8 N-terminal amino acids by cleaving the bond between gln 32 and asparagine, the N-terminus of the mature subunit[73]. The 36 kDa monomer is then assembled into a trimer in the mitochondrial matrix assisted by mhsp60, cpn10 and ATP. The bacterial homolog of mhsp60, GroEL, and of cpn10, GroES, are also able to chaperone the assembly of OTC monomers into active trimers[74] and require ATP to assemble the active trimer. MPP requires an intermediate octapeptide between the MPP cleavage site and the mature OTC subunit amino terminus[73]. MIP then requires an amino-terminal hydrophobic residue and a small residue at position 4 (ser, thr, gly) in the C-terminal octapeptide in order to recognize the cleavage site at the carboxyl side of glutamine 32. Pre-OTC is one of the best-studied examples of a number of twice cleaved leader peptides among imported mitochondrial proteins[73].

The uptake and transport of pOTC is rapid enough so that it normally does not accumulate in the cytoplasm. However, when the temperature is dropped to 25°C or the inner membrane charge is impaired by uncouplers of oxidative phosphorylation, then the intermediate form of pOTC accumulates in the inter-mitochondrial space[71]. The half-life of pOTC in the cytoplasm is under two minutes. An inhibitor of oxidative phosphorylation at the level of the FI-ATPase such as oligomycin or an inhibitor of ATP/ADP transport such as atractyloside does not block this transport. The combination of rotenone and antimycin-A that inhibit electron transport sites one and two respectively do block transport[(71)].

The OTC leader peptide has served as one of the most thoroughly studied models for uptake of cytosolic proteins into the mitochondria. The sequence of the human 32-amino acid peptide is M_1LFNLRILLNNAAFRNGHNFMVRNFRCGQPLQ_{32}. Counting from the N-terminal methionine, the leader peptide contains four arginine residues at position 6, 15, 23 and 26. Leader peptides have been synthesized making deletions of various parts of the peptide and substituting for various of the arginines. The conclusions of these studies are that the mid-portion of the molecule from residues 8 to 22 is essential for transport[(75,80)]. The critical amino acid seems to be arginine 23, which if replaced by glycine prevents normal cleavage. If arg 23 is replaced by lysine, pOTC is transported normally and converted to the 36 kDa mature subunit. If alanine is substituted for arg 23, 45 percent of the normal subunit results and if asparagine is substituted, the 36kDa subunit is only 20% of normal[(79)]. This study suggests that the positive charge at position 23 is the most critical element.

Leader peptides with deletions of amino acids 2 and 3 are converted to normal subunits better than those with deletions 2 to 7 and the latter better than those with deletions 2 to 12. A construct with deletions 26 to 31 is also converted. If glycine is substituted for arg 6 there is 80 percent uptake; if glycine is substituted for arg 15 there is 40 percent conversion; if glycine is substituted for arg 26 there is a 100 percent conversion. If both arg 15 and 26 are substituted there is 15 percent conversion and none if glycine 23 is present. The construct in which glycine is at position 23 cannot be cleaved by MPP normally but is cleaved at the 16-17 bond adjacent to arg 15[(72)] This glycine 23, 17-N-terminal derivative accumulates in the matrix space uncleaved by MIP at the gln 32-asn 1 bond, unable to form trimers and eventually is degraded. MIP, a zinc protease, is inhibited as expected by 1,10 phenanthroline, EDTA and by sulfhydryl agents. MIP will only cleave the intermediate form of OTC. In 28 proteins that are transported from the cytosol into mitochondria and have a leader peptide, the first cleavage is usually carried out by a general neutral protease of the inner membrane (MPP)[70]. The cleavage site is 0 to 2 residues on the carboxyl side of an arginine. The idea that an alpha helix is necessary in the center of the molecule has been questioned by studies that show that the glycine 23 derivative, a helix-breaking substitution, is imported normally into mitochondria where it is cleaved erroneously[79] by MPP.

Shore and colleagues have studied the physical-chemical properties of OTC leader peptides. A synthetic 1-27 OTC peptide binds to anionic phospholipids and lyses membranes to disc-like micelles[81] and competitively prevents pOTC uptake into mitochondria[82]. When the leucines at positions 5, 8 and 9 are replaced by alanine, the rate of import is reduced 5-fold but the precursor is processed to 36 kDa subunit normally[83]. When the pOTC leader peptide is linked to dihydrofolate reductase (pO-DHFR) the construct binds and is transported through the outer and inner membranes at contact points, if ATP and reticulocyte lysate protein are present[84]. This pO-DHFR uptake requires a cytosolic protein that contains a reactive thiol group and a 70 kDa heat shock protein which retards precursor folding and aggregation. Cote did a computer analysis of the OTC leader peptide and showed that the N-terminal region l-20 comprises a helical hydrophobic patch opposite to a hydrophilic surface carrying basic residues at potentially strategic positions[86]. When added *in vitro,* a synthetic peptide 1-19 of the rat precursor blocked the import of OTC. Once formed, trimeric OTC is loosely bound to the inner mitochondrial membrane[87,88] where it can participate in channeling substrate from the ornithine transporter and carbamyl phosphate from CPS-1.

Although nonstructured in an aqueous environment, a 32-residue synthetic rat OTC leader peptide developed a helix nucleation site over the hydrophobic cluster I7-L8-L9 when placed in 10% trifluoroethanol (TFE) to mimic a lipid environment[709]. As the TFE concentration was increased to 30%, the helical region expanded to include N4 to K16. The equilibrium constant between helix and random coil formation showed a strong temperature dependence with maximal values between 20 and 30 °C. 30% TFE placed the positively charged residues R6, K11, R15 and K16 on one side of the helix where they may interact electrostatically with negative charges on the Tom proteins in the outer membrane[709].

Commentary. pOTC is synthesized on free ribosomes, moves to the outer mitochondrial membrane, bound to chaperone proteins such as

MSF, hsc70, HSDJ and a presequence binding factor (PBF) which keeps pOTC in an import competent form. This transport complex passes pOTC off to binding sites on the outer membrane while cleaving ATP. It passes through a channel in the outer membrane with the help of Tom proteins and through a channel in the inner membrane with the help of Tim proteins, mhsp70, mGrpE and ATP. The electrochemical charge on the inner membrane is essential for transport. On the matrix side of the inner membrane MPP, a neutral protease, cleaves 24 N-terminal amino acids, leaving a phenylalanine as the new N-terminus of an octapeptide leader. MIP, a zinc protease, then cleaves the glutamine-asparagine bond that generates the N-terminus of the mature OTC subunit. Trimeric and enzymatically active OTC is assembled, loosely attached to the inner membranes, by mitochondrial hsp60, chaperonin 10 and ATP. OTC is a classic example of a mitochondrial protein whose precursor leader peptide is twice cleaved to become a mature subunit.

3

Molecular and Kinetic Characteristics

Ornithine transcarbamylase has been isolated and purified partially or extensively from five mammalian livers: human[89–91], rat[93–96], beef,[97–99], pig[100] and dolphin *(Stenella)*[101] where OTC serves for detoxification of ammonia by the urea (ornithine) cycle. Urea synthesis and OTC activity occurs in a wide spectrum of vertebrate species for various reasons: amphibians after metamorphosis from the ammoniatelic state[102,103]; turtles and tortoises for ammonia detoxification[103,104]; sharks, rays, skates, *Chimaera* and coelacanths using urea as an osmolyte[104,105]; a gobiid fish *(Mugilogobius abei)*[710] and an estivating lungfish for ammonia detoxification[106]; toadfish using urea for nitrogen cycling between liver and gut[104]; and for ammonia detoxification some teleost fish embryos and larvae (rainbow trout, cod[107]), freshwater, air-breathing teleosts[108] and a tilapia fish living in an alkali lake[109]. OTC activity has been detected in many species of mammals in addition to those noted above[102]. Some hibernating mammals use the urea cycle to recycle nitrogen between liver and gut[104]. Although birds excrete waste nitrogen as uric acid, the chicken kidney contains OTC for unknown reasons[110]. Non-vertebrates which have a functioning urea cycle and OTC include earthworms and land planarians[111]. A number of higher plant species as well as blue-green algae[112] contain OTC. Anong the OTCs highly purified from plants are those from pea seedlings *(Pisum sativum)*[112–114], *Citrus limonum* leaves[115], kidney bean leaves *(Phaseolus vul-*

garis)[116], crucifer leaves *(Arabidopsis thaliana)*[117] and legume leaves *(Canavalia lineata)*[118]. Eubacteria, archaeabacteria and yeasts and fungi use OTC as part of the arginine biosynthetic pathway (anabolic OTCs), in some cases as part of the arginine deiminase pathway (catabolic OTCs) and rarely as one protein with anabolic and catabolic functions depending on pH and effectors. We expect OTC to be present in all bacterial species.

In Table 3-1 we have tabulated the molecular weights of *purified* OTCs from many species and their subunit sizes when known. Other bacteria whose OTC molecular weights were determined on crude extracts are: *Lactobacilllus fermenti* (anabolic 125 kDa, catabolic 420 kDa), *Proteus vulgaris* (anabolic 125kDa), *Pichia fermentans* (anabolic 125 kDa), *Staphylococcus aureus* (?catabolic 200 kDa)[126] and *Streptococcus* D10 (?catabolic kDa 220)[131].

All subunits exist in the size range of 35–44 kDA (Table 3-1). In general, anabolic OTCs function as trimers with the notable exeptions of the two actinomycete OTCs whose hexameric proteins can function as anabolic or catabolic enzymes and *P.furiosus* OTC whose dodecamer has only anabolic function. Among bacteria, anabolic *A.formicans* OTC at pH 6 has a molecular weight by molecular sieving of 360 kDa (a nonamer) but at pH 8.5 dissociates to a trimer of 120 kDA[132]. Catabolic *A.formicans* OTC at pH 6.5 has a kDa of 420 and at 8.5 it is 125 and both are active[126]. *Str.*faecalis has an intermediary OTC kDa of 223 and functions both anabolically and catabolically depending on the pH and substrate concentrations[133] (Table 3-2).

In Table 3-2 we have listed for 34 species the specific activities, pH optima and the Michaelis constants (K_ms) for ornithine, carbamyl phosphate, citrulline and phosphate, or arsenate which is used in the reverse reaction of arsenolysis of citrulline. We only listed those species where completely or partially purified OTCs were used to measure the kinetics. In some catabolic OTCs the substrate concentrations at half maximal velocity ($S_{0.5}$) are given along with the Hill coefficients (n_H). The inhibition constant (K_i) for phosphate versus CP is reported where a K_m for phosphate in the reverse reaction was not measured. In mammalian livers the K_ms for ornithine in the anabolic, forward reaction range from 0.14 to 0.40 μM but vary much more in other species. The K_ms for carbamyl phosphate in the five mammals ranged from 0.06 to 1.3mM. The specific activities of the mammalian

Table 3-1. Molecular weights and subunit weights of OTCs

Sources	Native OTC	Anabolic (a) or catabolic (c)	Subunit of OTC
Mammals			
Human liver	110[90]	a	38[91]
Rat liver[95]	112	a	40
Beef liver[99]	108	a	37.8
Pig liver[100]	107	a	36.8
Dolphin liver[101]	103	a	34.4
Birds			
Chicken kidney[110]	110	a	36
Sharks			
Squalus acanthias[119]	114	a	38
Sphyrna zygaena[101]	108	a	36.4
Plants			
Pea seedling, *Pisum sativum*[114]	108	a	36.5
Kidney bean leaves, *Phaseolus vulgaris*[116]	109	a	36
Crucifer leaves, *Arabidopsis thaliana*[117]		a	37
Legume leaves, *Canavalia lineata*[118]	107	a	38
Yeasts, fungi			
Saccharomyces cerevisiae[120]	110	a	37
Neurospora crassa[121]	110	a	37
Actinomycetes			
Streptomyces clavuligerus[122]	248	a,c	37
Nocardia lactamdurans[122]	251	a,c	37
Archaeabacteria			
Halobacterium halobium (salinarium)[43]	200	c	41
Pyrococcus furiosus[123]	400	a	35
Eubacteria			
Thermus thermophilus[48]	110	a	36
Escherichia coli W[124]	105	a	35
Pseudomonas fluorescens[125]	110	a	
" "[125]	420	c	
Pseudomonas aeruginosa[35]	110	a	36
" "[34]	420	c	40
Pseudomonas putida[126]	110	a	
" "[126]	420	c	
Aeromonas formicans[132]	120[360]	a	36
" "[126]	420[125]	c	41.5

Sources	Native OTC	Anabolic (a) or catabolic (c)	Subunit of OTC
Bacillus licheniformis[126]	140	a	44
Bacillus licheniformis[126]	140	c	44
Streptococcus faecalis[99]	223	a,c	39.6
Bacillus subtilis[127]	260	a	44
Salmonella typhimurium[128]	116	a	39
Mycobacterium bovis[41]	110	a	38
Mycobacterium smegmatis[129]	116	a	
Mycoplasma hominis[130]	360	c	

Note: Molecular weights of OTCs determined by ultracentrifugation, gel filtration or sucrose density gradient centrifugation; subunit weights by SDS-polyacrylamide electrophoresis.

OTCs vary greatly. The low value for the pig liver OTC[100] may partly be due to use of Tris buffer which long ago was shown to inhibit beef[98] and human[141] OTCs and to lower the apparent pH optima.

The K_m for CP in *T.thermophilus* was 0.1mM and in *Pisum sativum* was 0.2mM even though the binding site for CP in the first was SLRT and was SMRT in the latter rather than the ST(59)RT which is found in all but 4 of 33 OTCs whose amino acids have been sequenced[48]. Thus T59 (human sequence) does not play a unique role in CP binding as pointed out by Sanchez et al.[48]. In the actinomycetes one hexameric enzyme functions in both anabolic and catabolic directions and the same double function occurs in hexameric *Str. faecalis* OTC controlled by pH and substrate concentrations (Table 3-2). The anabolic function peaks at pH 8.4 at a high velocity while the catabolic reaction, phosphorolysis of citrulline, peaks at pH 7.1 at a low rate and at a high K_m for citrulline[133].

In other eubacteria two different OTCs are programmed by separate genes, with the larger catabolic enzymes showing allosteric, cooperative kinetics (i.e., *Ps.aeruginosa)* and the trimeric anabolic enzymes showing Michaelis-Menten kinetics. The substitution of alanine for glutamate at position 105 (E105A) in the allosteric *Ps. aeruginosa* OTC abolishes the cooperativity of the wild-type enzyme for CP and makes the response to CP hyperbolic at all pHs[140]. In the wild-type enzyme 10 mM phosphate also shifts the response to CP from a sigmoidal to a hyperbolic curve, so phosphate serves both as an activator up to 10 mM and a product inhibitor competitive with CP[140].

Table 3-2 Michaelis constants, specific activities and pH optima of OTCs K_ms (mM) at 37°C and pH 7.7–8.5

Sources	Anabolic (a) or catabolic (c)	Ornithine	Carbamyl phosphate	Citrulline	Phosphate (arsenate)	Specific activity[a]	pH optimum
Mammals							
Human liver	a	0.40[89]	0.29[89]			245[91]	7.7[89,91]
" "	a	0.40[91]	0.16[91]				
" "	a	0.20[90]	0.09[90]				
Rat liver[95]	a	0.4	0.26		0.25K_i	885	7.6–8.1
Beef liver[133]	a	0.14	0.4			780	8.5
Pig liver[100]	a	0.41	1.3			26.4	8
Dolphin liver[101]	a	0.14	0.062			211	7.8
Birds							
Chicken kidney[110]	a	1.21	0.11			77.1	8.5
Amphibians							
Frog liver[134]	a	1.5	2.9		4.9K_i	146	8.3
Shark livers							
Squalus acanthias[119]	a	0.71	0.053			27	7.8
Sphyrna zygaena[101]	a	1	0.083			115	7.8
Scylliorhinum caniculum[134]	a	1.7	0.42		1.82	137	8.1

Sources	Anabolic (a) or catabolic (c)	Ornithine	Carbamyl phosphate	Citrulline	Phosphate (arsenate)	Specific activity[a]	pH optimum
Plants							
Pisum sativum[112]	a	1.2	0.2		1.2K_i	139[114]	8.5[114]
Citrus limonum leaves[115]	a	0.7	0.1		4K_i		9
Phaseolus vulgaris[116]	a	0.63	0.16			723	8
Canavalia lineata[118]	a	2.4	0.21			24	8.5
Yeasts, fungi							
Sacch. cerevisiae[116]	a	0.9	0.2	9[135]	3.4[135]	785	8.5
Neurospora crassus[117]	a	5	2.5			251	9.5
Actinomycetes							
Streptomyces clavuligerus[122]	a,c	2.3	6.1	1.37			8
Nocardia lactamdurans[122]	a,c	0.14	3.5				8
Archaeabacteria							
Halobact.halobium salinarium[43]	c	8	0.4				8.8
Pyrococcus furiosus[123]	a	0.13	0.13				6.5
Eubacteria							
Thermus thermophilus[48]	a	0.1	0.1			100(55°C)	8
E. coli W[124]	a	2.4	0.2	10	1.8	2830	8.5
E. coli K-12[136]	a	0.59	0.026			1880	8.3
Pseudomonas fluorescens[125]	c		$S_{0.5}$ 38[b] n_H 5.8[b]		10	753	7.8

Table 3-2 (continued)

Sources	Anabolic (a) or catabolic (c)	Ornithine	Carbamyl phosphate	Citrulline	Phosphate (arsenate)	Specific activity[a]	pH optimum
Pseudomonas fluorescens[125]	a		0.3			40	9
Pseudomonas aeruginosa	c	0.62[140]	$S_{0.5}$ 3.2[140]	$S_{0.5}$ 22[138]	$S_{0.5}$ 21[138]	225[139]	6.5[140]
			n_H 3.8b	n_H 2.8[b]	n_H 3.6[b]		
Strept. Faecalis[133]	*a*	0.68	0.29	26	1.9	3390	8.4
	c	1.1	45	17	0.8	89	7.1
Bacillus subtilis[127]	*a*	5	0.9			347	8.3
Salmonella typhimurium[128]	a	0.2	0.06			2490	8
Mycobacterium smegmatis[129]	a	0.25	0.2		0.82K_i	88	8.5
Streptococcus lactis[137]	a	2	3.7			1440	8.3
Bacillus licheniformis[126]	c					2020	8
Aeromonas formicans[126]	c	0.9	0.5			1380	8
Aeromonas formicans	a	20[126]	0.2[126]			3820[132]	8.5[132]
Mycoplasma hominis[130]	*c*	3.6	2.3	0.4		6300	8.4

a. Specific activities expressed as μmol citrulline/min per mg of purified enzyme with assay temperatures all normalized to 37°C with a Q_{10} of 1.87.

b. $S_{0.5}$ = substrate concentration at 0.5 maximum velocity; n_H = Hill coefficient of the sigmoidal saturation curve.

Ornithine has three molecular dissociation constants, 1.705, 8.69 and 10.755[142]. The neutral form of ornithine has been shown to be the substrate of ornithine transcarbamylase[89,133]. In the *E.coli* K12 arg I enzyme the form of neutral ornithine which is the actual substrate is the minor zwitterionic species H_2N $(CH_2)_3$ CH $(NH_3+)COO-$ which has a pK of 9.8 for the alpha-amino group and accounts for only 17% of the ornithine species which have zero net charge[136]. Thus, when the concentration of the neutral form of ornithine at each pH is calculated, the Km for neutral ornithine can be derived. For the human enzyme this averages to be 0.039 mM[89], for beef liver OTC 0.026 mM[133] and for E. coli K-12 OTC 0.064 mM[136]. The Km for the actual zwitterionic substrate would be 17% of these values.

The reverse reaction of OTC can be accomplished by using arsenate instead of phosphate whereby the carbamyl arsenate formed is broken down spontaneously to CO2 and ammonia. The Km's for citrulline and phosphate in mammalian OTCs have not been determined in pure enzymes. When Krebs defined OTC as a specific mammalian enzyme[143] he determined a Km (citrulline) of 14 mM in guinea pig liver homogenates. In Table 3-2 the non-mammalian K_ms (citrulline) vary from 1.4 to 26 mM, and the K_ms(arsenate) from 0.8 to 10 mM. Beef liver OTC also has the ability to decarbamoylate carbamoyl-hemoglobin and carbamoyl histones[144].

The pH optimum for the forward reaction differs among these enzymes: mammalian liver OTCs peak at 7.6–8.5 at 37°C. The other anabolic OTCs in Table 3-2 have pH optima from 7.8 to 9.5 except for *Pyrococcus furiosus* which peaks at pH 6.5[123]. The pH optima for the catabolic enzymes range from 6.5 to 8.8 although the reverse reaction peaks at pH 7.1 in mammalian OTCs. The isoelectric point (pI) values for mammalian OTCs range from 6.8[91] to 8.0[90] for the human to 8.1 for the rat OTC[93], 6.95 for beef OTC[99] and 8.6 for the dolphin OTC[101]. By deriving Arrhenius plots the activation energy or enthalpy has been reported as 13.3 Kcal/mole for the beef enzyme[133] and 11 Kcal/mole for E. coli K-12 OTC[136]. Marshall[133] determined the equilibrium constant for the beef liver OTC as 6×10^5 M at pH 7.9, which indicates how far toward citrulline the equilibrium lies. The maximum velocity in the reverse reaction of beef OTC is only l/40 of that in the forward reaction at optimal pHs and substrate concentrations: Vf/Eo = 3580(S^{-1}) at pH 8.8 and Vr/Eo = 89(S^{-1}) at pH

7.1[133]. The relative half-life of rat liver OTC is 6–9 days, similar to that of CPS, and of mitochondria[145].

An inhibitor of the human enzyme is norvaline (2-aminopentanoic acid) with a K_i of 0.24mM[90], almost identical to the K_m for ornithine with which it competes. The delta-amino group in ornithine (2,5-diamino-pentanoic acid) is replaced in norvaline by a methyl group, which indicates that the neutral form of ornithine which binds to the enzyme is the form with the alpha or 2-amino group charged and the delta or 5-amino group uncharged. 2-amino-4-pentenoic acid is also an inhibitor with a K_m of 0.47mM[90]. This is an analog of the toxin hypoglycin that is known to cause Jamaican Vomiting Disease and hyperammonemia. Lysine (2,6-diaminohexanoic acid) is a weakly bound substrate yielding homocitrulline and is a very weak inhibitor which shows that the length of the side chain is more important at the ornithine binding site than the presence of an uncharged amino group on the end of the side chain[90]. The most potent inhibitor of OTC is N^{δ}-(phosphonacetyl)-L-ornithine (PALO), a transition state analogue, which resembles structurally the carbamyl phosphate-ornithine complex. It is competitive with carbamyl phosphate with a K_i of 15 μM at pH 7.2 and noncompetitive with ornithine[146].

In the human enzyme the neutral form of ornithine accounts for the inhibition that occurs at high concentrations of ornithine[89]. This inhibition grows stronger as the pH increases and more neutral ornithine is generated. When the neutral ornithine concentration is greater than 0.25 mM inhibition commences. Therefore, all kinetic studies involving ornithine should use a concentration at a given pH that is not inhibitory but this problem is ignored in many kinetic studies. When the maximal velocity derived from double-reciprocal plots was obtained at various pHs, a sigmoidal curve resulted which fits the ionization of a single group on the enzyme with an inflection point at pH 6.55[89]. When carbamyl phosphate was varied the inflection point was at pH 6.8. Because no groups on ornithine ionize at this pH and the K_ms for carbamyl phosphate and neutral ornithine did not change with pH it was postulated that the ionizing group must be on the enzyme. From the pKs of 6.6–6.8 this group that must ionize in order to derive maximal activity should be the imidazolium nitrogen of a histidine residue[89]. This has been subsequently confirmed by a histidine site-specific reagent[147] and by an experiment of nature where replacement of histidine-85 by asparagine in the mature monomer of the sparse fur mouse OTC reduces activity at pH 7.7 to 20% of normal, shifts the pH optimum from 7.7 to 9, moves the inflection point of the V_{max} vs. pH curve to 8.7, impairs ornithine binding and causes hyperammonemia[148,149].

Norvaline is a competitive inhibitor versus ornithine in the rat liver OTC but the Ki of 0.07 mM[95] is much lower than that for the human OTC. Leucine is also a competitive inhibitor with a Ki of 1.2 mM. Lysine serves as a substrate at pH 8.8, forming homocitrulline. The Km for lysine is 40 mM, a value so high that lysine inhibition of OTC is unlikely to occur. No inhibition of the rat liver OTC by orotate or arginine occurs[95], although some bacterial OTCs show feedback inhibition by these compounds which occur in the pyrimidine and arginine biosynthetic pathways, respectively.

Because the concentration of carbamyl phosphate probably does not exceed 18μM[150] in rat liver mitochondria and the K_m value for carbamyl phosphate in rat liver is around 0.26 mM, this means that OTC is not saturated in the mitochondria and carbamyl phosphate should play a major role in controlling the rate of OTC activity *in vivo*. However OTC seems to prefer endogenously formed CP in mitochondria, not the CP added to the incubation medium[151]. The level of carbamyl phosphate in mitochondria is controlled by carbamyl phosphate synthetase-1 whose activity is governed by the mitochondrial free ammonia concentration and the amount of N-acetylglutamate relative to the amount of CPS-1[151].

The concentration of ornithine can vary in rat liver mitochondria on 5–70% casein diets from 0.5 to 1.4 mM[152] and in isolated hepatocytes is 1.6 mM in the mitochondrial matrix compared with 0.6 mM in the cytosol[153]. The K_m for total ornithine at pH 7.4 in human mitochondria averages approximately 0.8 mM[89] but there is evidence for channeling of ornithine from its carrier on the mitochondrial membrane to OTC and OTC uses extramitochondrial ornithine in preference to matrix ornithine[154]. The Km for the external side of the mitochondrial carrier for ornithine is 0.16 mM[155] and the ornithine/citrulline carrier catalyses an electroneutral exchange of ornithine for citrulline plus an H^+ generated by the OTC reaction. Thus the cytoplasmic, not the mitochondrial ornithine concentration may play a major role in controlling the minute-to-minute rate of citrulline and therefore urea synthesis.

Beef liver enzyme kinetics are compatible[133] with an ordered, sequential reaction in which carbamyl phosphate binds first to the enzyme and then ornithine binds to the enzyme-carbamyl phosphate complex, followed by the catalytic event of carbamyl transfer which gives an enzyme-citrulline-phosphate complex. Citrulline dissociates first and then phosphate dissociates from the enzyme. This dissociation of phosphate is considered by some to be the rate-limiting step in the forward reaction. The mechanism by which ornithine inhibits the forward reaction has been studied both by Marshall(133) in the beef liver OTC and by Legrain and Stalon(124) in the *E.coli* enzyme. In *E. coli* W OTC, ornithine inhibition is said to occur by two mechanisms. First, neutral ornithine binds weakly to the free enzyme resulting in non-competitive inhibition versus carbamyl phosphate, which is the preferred first binding substrate. Secondly, phosphate binds to the enzyme and then this complex binds neutral ornithine forming a dead-end complex of enzyme-phosphate-ornithine that results in non-competitive inhibition versus carbamyl phosphate. As the pH rises the concentration of neutral ornithine increases and the first mechanism is more active. The second mechanism requires phosphate and occurs even at acid pHs. If the ornithine concentration exceeded 1 mM within mitochondria we calculate that human OTC could be inhibited by this second mechanism in the presence of 10–20 mM phosphate at pH 7.4 in the mitochondria.

Kuo has reported that micromolar, physiologic concentrations of Zn^{2+} bind competitively with ornithine inducing an allosteric positive cooperativity upon the saturation curve for ornithine in the E. coli K-12 OTC(156). The usual hyperbolic response of activity to ornithine becomes sigmoidal, with a limiting Hill coefficient of 2.7 at 0.3 mM Zn^{2+}. When Zn^{2+} binds to Cys 273 of E. coli OTC(157), it induces slow isomerization of the enzyme and acts as an inhibitor. When OTC is first bound with carbamyl phosphate, zinc binds cooperatively to the binary complex with a Hill coefficient of 1.5(158). Zn^{2+} then is competitive against ornithine, non-competitive against carbamyl phosphate and induces no enzyme isomerization. Neither Co^{2+} nor Cu^{2+} exert this Zn^{2+} effect on the enzyme. The role of Zn^{2+} in modulating OTC activity *in vivo* is not clear; allosteric kinetics are not found in the E. coli K-12 enzyme as isolated, and the role of Zn^{2+} in mammalian OTCs has not been studied.

Commentary. OTC has been found and often purified in a wide spectrum of life forms across the evolutionary scale. In eubacteria, archaeabacteria, yeasts, fungi and plants OTC usually functions in the arginine biosynthetic pathway but in some bacteria it serves in an arginine degradation pathway. In mammals and some other vertebrates OTC functions as part of the urea/ornithine cycle to detoxify ammonia, but in others it makes urea as an osmolyte or as a way to cycle nitrogen between liver and gut. The subunit size in all species is 35–44 kDa. The anabolic OTCs usually function as trimers with hyperbolic substrate responses. The catabolic OTCs are usually polymers of 6,9 or 12 subunits that show sigmoidal substrate curves. Kinetic parameters are similar in most anabolic OTCs. The form of ornithine that is the actual substrate is one of the minor neutral species. Ornithine seems to be channeled by the ornithine/citrulline carrier on mitochondrial membranes to OTC loosely bound to the inner membrane in the matrix.

4

Active Site and Other Essential Residues

With the collaboration of Thomas Hurley, PhD

Before the three-dimensional structure of an OTC molecule had been determined, we had to extrapolate its probable structure from that of *E.coli* aspartate transcarbamylase (ecoATC) whose structure and residues which bind carbamyl phosphate (CP) and aspartate had been delineated by Lipscomb and coworkers in a series of landmark X-ray diffraction studies ([159–164]), by Jin et al.[165] and by Endrizzi et al.[166]. *E.coli* ATC consists of two catalytic trimers and three regulatory dimers, the catalytic monomers being 34kD in size, each with a regulatory subunit 17kD in size[159]. ATC and OTC are closely related in the evolutionary scale but ecoOTC or other OTCs do not contain regulatory subunits. The catalytic monomer of ATC is very similar in size and linear sequence to that in OTC especially in the N-terminal, so-called polar domain where CP is bound. The C-terminal or equatorial domain in ecoATC includes residues 150–284. In ecoATC the domains are bridged by two interdomain α-helices, helix $\propto 5$ (135–149) that starts in the CP and ends in the aspartate domain and helix $\propto 11$ (285–304) that begins in the aspartate and passes through the CP domain.

In Fig. 4-1 we have displayed the linear sequences of human OTC (hOTC), mouse OTC, catabolic *Pseudomonas aeruginosa* OTC (paeOTC), *E.coli argI* OTC (ecoOTC) and *E.coli pyrB* ATC (ecoATC) in order to compare hOTC with the two bacterial OTCs which have had their 3-dimensional structure determined by X-ray diffraction without[167,168] and with a bound substrate analogue[169], and with mouse OTC because the two spontaneous models of OTC deficiency occur in the mouse (Chapter 6). The amino acids are aligned in the most likely matching positions based on the secondary structures of the OTC monomers reported by Villeret et al. for catabolic paeOTC[167], by Ha et al. in ecoOTC[169], by Tuchman et al.[170] and by Shi et al.[171]in hOTC, considering the differing sizes among the three monomeric subunits. Our alignment matches that of Labedan et al.[172,173] who used various computer and manual methods to align 33 OTCs and 31 ATCs from many species to construct a phylogenetic tree. Both paeOTC and ecoOTC contain an insertion of 19–20 amino acids in a row (277–296 in paeOTC and 277–295 in ecoOTC) comprising a loop-α-helix 9α-loop, not found in hOTC, mouse OTC or ecoATC in the C-terminal region of their monomers (Fig. 4-1). The hOTC and mouse OTC as well as other ureotelic OTCs contain a C-terminal extension (314–322) which folds back from the CP domain toward the ornithine domain and forms a ridge on the convex side of the trimer[171]. Shi et al. speculate that it mediates interactions with the mitochondrial membrane[171].

We used the 3-dimensional structures of ecoOTC(argI)[169] and paeOTC[167] whose atomic coordinates were then published in the Protein Data Bank, Biology Department, Brookhaven National Laboratory, to develop a computer model of hOTC (Modeller Release 4)[174], which predicted the actual structure of hOTC subsequently reported by Shi et al.[171] very closely, except for the C-terminal extension which does not exist in the bacterial sequences (our model not shown). We used the hOTC sequence of Horwich et al.[6] in Fig. 4-1 although Hata et al.[8] found these 5 polymorphisms compared to Horwich: F69L, P79L, C161W, F162I and R238Q. In Fig. 4-1 we have numbered the secondary structure elements as did Shi et al.[171] in hOTC including 14 α-helices and 10 β-strands per monomer. The extra loop a9a found in bacterial OTCs, residues 277–296 in ecoOTC and residues 277–295 in paeOTC, is seen in Fig. 4-1. This loop is also absent in ecoATC where it would presumably disrupt the binding of a catalytic to a regulatory chain[169]. Its function in ecoOTC and paeOTC is unknown; Ha et al. speculate that it interacts with the 240s and 80s loops[169]. In Fig. 4-2 we have reproduced from Shi et al.[171] a ribbon diagram of the hOTC monomer, with PALO bound. On the left is the N-terminal domain (residues 1–134) that binds CP.

```
Human OTC       1 NKVQLKG RDL LTLKNFT GEEIKYMLWLSADLKFRIKQK GEYLPLLQG KSLGMIF   54
Mouse OTC       1 SQVQLKG RDL LTLKNFT GEEIQYMLWLSADLKFRIKQK GEYLPLLQG KSLGMIF   54
Ps·aerug·OTC    1  AFNMHN RNL LSLMHHS TRELRYLLDLSRDLKRA·KYT GTEQQHLKR KNIALIF   52
E·coli OTC      1   SGFYH KHF LKLLDFT PAELNSLLQLAAKLKAD·KKS GKEEAKLTG KNIALIF   51
E·coli ATC      1  ANPLYQ KHI ISINDLS RDDLNLVLATAAKLKA····· NPQPELLKH KVIASCF   48
                          β1   α1a              α1                      β2

     o  o ooo                           o
55  EKRS TRTRLSTETGFALL G GHPCFPT·TQDIHLGVN·ES LTDTARVLSSMADAVLA   108
55  EKRS TRTRLSTETGFALL G GHPSFLT·TQDIHLGVN·ES LTDTARVLSSMTDAVLA   108
53  EKTS TRTRCAFEVAAYDQ G ANVTYID·PNSSQIGHK·ES MKDTARVLGRMYDAIEY   106
52  EKDS TRTRCSFEVAAYDQ G ARVTYLG·PSGSQIGHK·ES IKDTARVLGRMYDGIQY   105
49  FEAS TRTRLSFETSMHRL G ASVVGFSDSANTSLGKKGET LADTISVISTYVDAIVM   104
              α2             β3                      α3        β4

    o                                 o   o
109 RVYK QSDLDTLAKEA S·I PIINGL SD··LYH PIQILADYLTLQEHY· SS·LKG   156
109 RVYK QSDLDTLAKEA S·I PIVNGL SD··LYH PIQILADYLTLQEHY· GS·LKG   156
107 RGFK QEIVEELAKFA G·V PVFNGL TD··EYH PTQMLADVLTMREHS·DKP LHD   155
106 RGYG QEIVETLAEYA S·V PVWNGL TN··EFH PTQLLADLLTMQEHLPGKA FNE   155
105 RHPQ EGAARLATEFS GNV PVLNAGDGS··NQH PTQTLLDLFTIQETE·GR · LDN   154
           α4              β5                   α5

                 o
157 LTLSCF GDG··N NILHSIMMSAAKF·GMHLQAATP KGY EPD ASVTKLAEQYAKEN G   209
157 LTLSWI GDG··N NILHSIMMSAAKF·GMHLQAATP KGY EPD PNIVKLAEQYAKEN G   209
156 ISYAYL GDAR·N NMGNSLLLIGAKL·GMDVRIAAP KAL WPH DEFVAQCKKFAEES G   209
156 MTLVYA GDAR·N NMGNSMLEAAALT·GLDLRLVAP QAC WPE AALVTECRALAQQN G   209
155 LHVAMV GDLK YGRTVHSLTQALAKFDGNRFYFIAP DAL AMPQYILDMLDEK····· G   206
      β6               α6             β7                α7
```

```
                                     o        o o
210 TKLLLT ND PLEAAH GGN VLITDTW   ISMGR·EEE KKKRLQAFQGY QVT MKTAKV ·AASD 265
210 TKLSMT ND PLEAAR GGN VLITDTW   ISMGQ·EDE KKKRLQAFQGY QVT MKTAKV ·AASD 265
210 AKLTLT ED PKEAVK GVD FVHTDVW   VSMGEPVEA WGERIKELLPY QVN MEIMKA TGNPR 267
210 GNITLT ED VAKGVE GAD FIYTDVW   VSMGEAKEK WAERIALLREY QVN SKMMQL TGNPE 267
207 IAWSLH SS IEEVMA EVD ILYMTRVQKERL DPS··  ··EYANVKAQF VLR ASDLHN AKANM 261
     β8        α8        β9                     α8a              α9
         o o                                                      o
266 WTFLH CLPR                        KPEEVD  DEVFY S PRS L VFPEAENRKW      300
266 WTFLH CLPR                        KPEEVD  DEVFY S PRS L VFPEAENRKW      300
268 AKFMH CLPAFHNSETKVGKQIAEQYPNLANGIEVT EDVFE S PYN I AFEQAENRMH      322
268 VKFLH CLPAFHDDQTTLGKKMAEEF·GLHGGMEVT DEVFE S AAS I VFDQAENRMH      321
262 ·KVLH PLPR                         VDEIA  TDVDK T PHA W YFQQAGNGIF      294
     β10              α9a                       α10           α11

301 TIMAVMVSLL TDYSPQLQKPKF                                              322
301 TIMAVMVSLL TDYSPVLQKPKF                                              322
323 TIKAILVSTL ADI                                                        335
322 TIKAVMVATL SK                                                         333
295 ARQALLALVL NRDLVL                                                     310
        α11
```

Figure 4-1. Amino acid sequences and secondary structure alignment of human, mouse, *Pseudomonas aeruginosa* and *E. coli* OTCs and *E. coli* aspartate transcarbamylase (ATC). Sequence alignments given for human OTC[6], mouse OTC[13], *Ps. aerug.* OTC[34,167], *E. coli* OTC (argI)[30,168] and *E. coli* ATC (pyrB)[31,172]. For human OTC, α-helices are double-underlined and β-strands are *italicized,* labeled as in references 167, 169, 171. Secondary structure of mouse OTC is made identical to hOTC. Residues in boldface are conserved in 33 OTCs and 31 ATCs[172,173]. Residues proposed to be involved in hOTC substrate binding or catalytically active sites as listed in Table 4–2 are marked by open circles above the amino acid.

On the right is the C-terminal domain (residues 152–290)that binds ornithine and undergoes a large conformation change when the substrate PALO binds. The first interdomain helix H5 ($\alpha 5$) uses residues 137–151 and the other interdomain helix H11 ($\alpha 11$) comprises residues 291–310. The interdomain cleft is the site where PALO or the two natural substrates enter, and is about 20 Å across and 15 Å deep. In Figure 4-3 we have reproduced from Shi et al.[171] a ribbon diagram of the OTC *trimer,* viewed down the molecular 3-fold axis from the convex face. In ecoOTC it forms a triangular structure about 54 Å

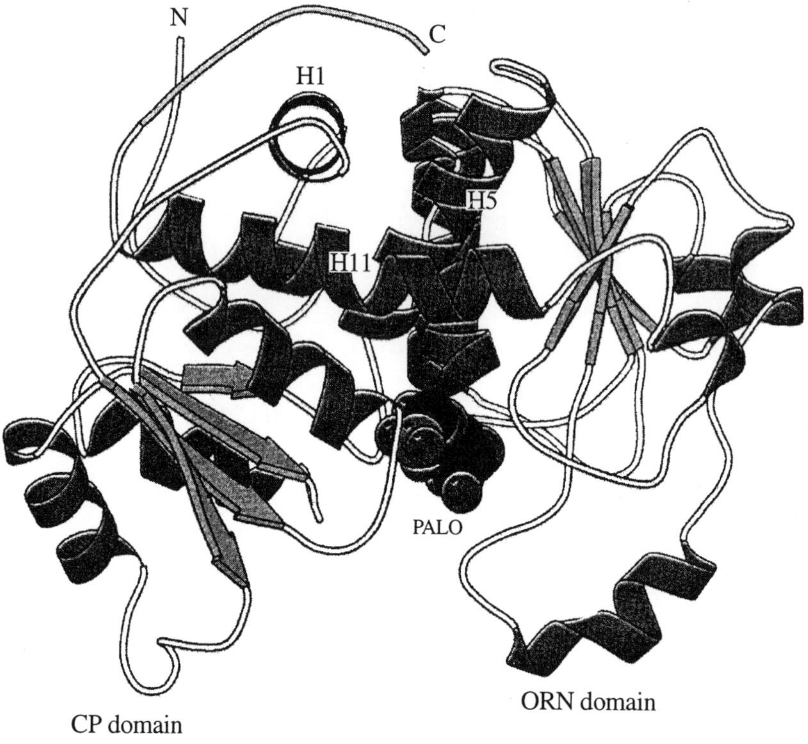

Figure 4-2. Ribbon diagram of human OTC monomer liganded with the bisubstrate analog PALO[171]. α-Helices are shown in black, β-strands in gray and random coils as strands. The C-terminal extension (residues 345–354), near helix H1($\alpha 1$), is shown as a gray strand. Helices H11($\alpha 11$) and H5($\alpha 5$) are interdomain helices that link the CP and ORN domains. The bisubstrate analog, PALO, is shown as a space-filling model. (Copyright 1998. Reprinted with permission of the Amer Soc Biochem & Molec Biol.)

deep and 92 Å wide with a molecular 3-fold axis of symmetry passing through its center[169]. From the orientation of Fig. 4-3 the three monomers of the trimer form a cup one-third the depth of the enzyme and the three active site clefts open into this cavity. The three N-terminal domains form the bottom of the cup and the C-terminal domains form the sides of the cup[169].

In Fig. 4-4, utilizing the above-mentioned ATC and OTC structural studies and models, we have drawn a secondary structure model of

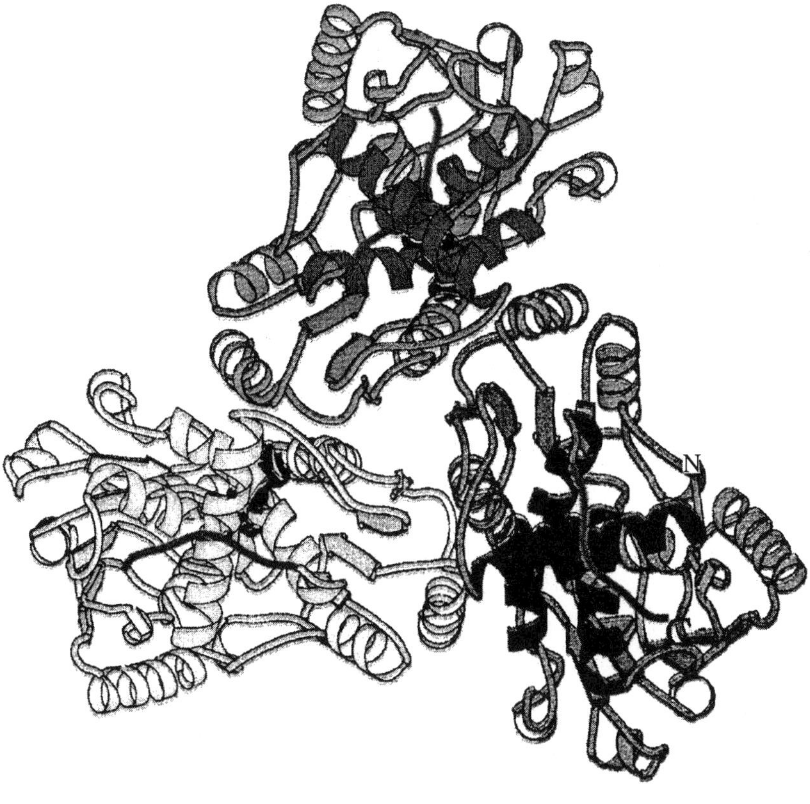

Figure 4-3. Ribbon diagram of the human OTC catalytic trimer[171]. The bisubstrate analog PALO, shown in black, interacts with residues from two adjacent subunits. The C-terminal extensions, shown in gray, are exposed on the convex face of the enzyme. The view is down the molecular 3-fold axis from the convex face. (Copyright 1998. Reprinted with permission of the Amer Soc Biochem & Molec Biol.)

the *human* OTC monomer. The α-helices are shown as cylinders and the β-strands as arrows, with the beginning and end of each numbered from the N-terminus. The polar domain is to the left and the equatorial domain to the right. Helix α1, using the nomenclature of Ha et al.[169], is positioned between the two domains. Shown in black circles are the essential substrate binding groups described in Table 4-2. The approximate location of the two substrates, CP and ornithine are shown labeled inside circles. The 80s loop on the left of the polar domain becomes part of the binding site for CP and ornithine for the

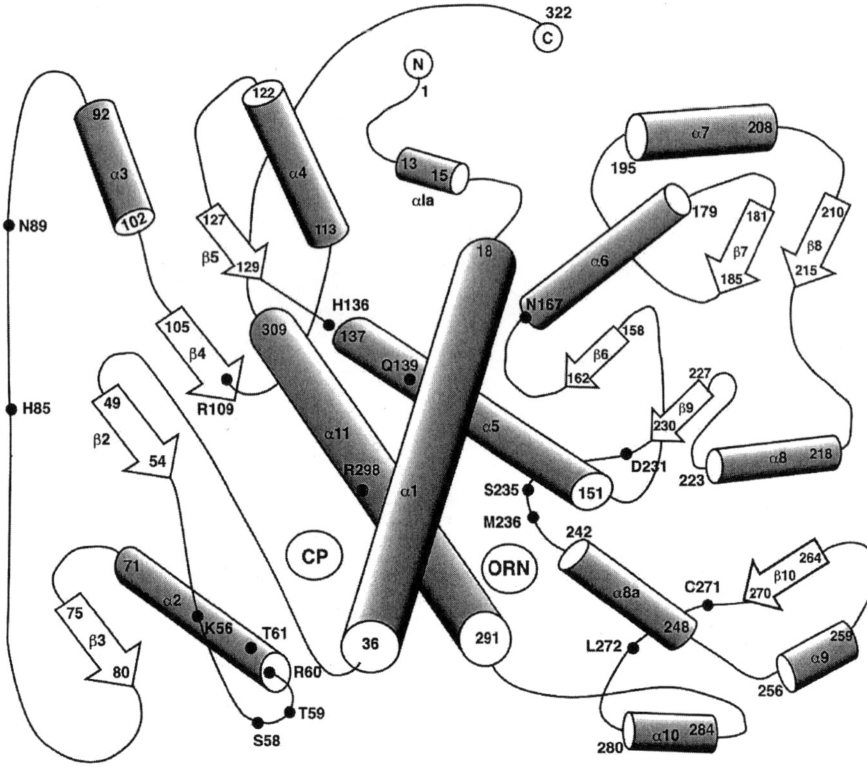

Figure 4-4. Structural model of the human OTC monomer. α-Helices are shown as cylinders and β-strands as arrows. The amino acid sequence numbers are placed at the beginning and end of each helix or strand. The substrate binding groups described in Table 4–2 are shown as black circles. The approximate location of the two substrates, CP and ORN, are shown in open circles.

monomer on the left whereas the 80s loop of the monomer below inserts into the cleft formed by helix ∝1 above and ∝2 and ∝11 below.

We have reproduced a figure of Shi et al.[171] showing the active site residues which bind PALO (Fig. 4-5), as a stereo view above and as a schematic below. The active site residues should be compared with Table 4-2.

The sequence of the OTC monomer has been amazingly conserved over the course of evolution[172]. Tuchman et al.[170] compared the primary sequences of 26 OTCs deposited in the Predict Protein server at the European Molecular Biology Laboratory in Heidelberg, Germany. Only M1, R6 and V22 are invariant in the N-terminal leader peptide of seven species where this leader is necessary for uptake into mitochondria. According to our review of the sequence matches of Labedan et al.[172,173] there are 30 *invariant* residues in 33 OTCs from bacteria, fungi, plants, amphibians and three mammals, mouse, rat and human (Table 4-1). Thirteen of 16 residues we listed in Table 4-2 as being involved in binding substrates or in catalysis are invariant and two are almost invariant. The sixteenth residue in Table 4-2, H85 in hOTC binds CP but is not invariant in other OTCs. Two more residues are invariant (L317 and F322) in those ureotelic OTCs whose C-terminal sequences are extended to 317 or 322 amino acids. Fifteen amino acids out of 33 species are *almost invariant,* defined as having only 1–4 residues differing from the most common amino acid in these 15 species lists (Table 4-1). Thirty six additional residues seem to be interchangeable among the 33 OTCs, presumably because they are homologous in structure, function and polarity, such as L-V-I (18 examples), R-K(3), T-S(4), G-A(5), W-Y-F(3), and E-D(3). These positions allow more variation than just these homologues, accepting 1–6 other amino acids while still maintaining function (Table 4-1). Little variation is apparent when the sequences of the only three mammalian OTCs, human, mouse and rat, are compared: 16 residues in the mouse and 19 in the rat differ from hOTC. None of these differences occur among the invariant, almost invariant or homologous residues listed in Table 4-1. Moreover none of the differences occurs at a residue where human missense mutations in the OTC gene have caused functional defects (see Table 5-2).

From our alignments in Fig. 4-1 we compared the similarities of the sequences of hOTC with ecoOTC and with paeOTC. One hundred twenty two or 39% of the 311 paired sequences of hOTC and ecoOTC are identical. If pairs that show functional homology, as listed above, are included, 31 more for a total of 153 or 49% are similar when 311 of the amino acids of hOTC and ecoOTC are paired as in Fig. 4-1. The bridging helix $\alpha 5$ of ecoOTC (135–148) has 10/15 amino acids identical with that of hOTC. The C-terminal bridging helix $\alpha 11$ (312–331) has 13/20 matched in hOTC and ecoOTC.

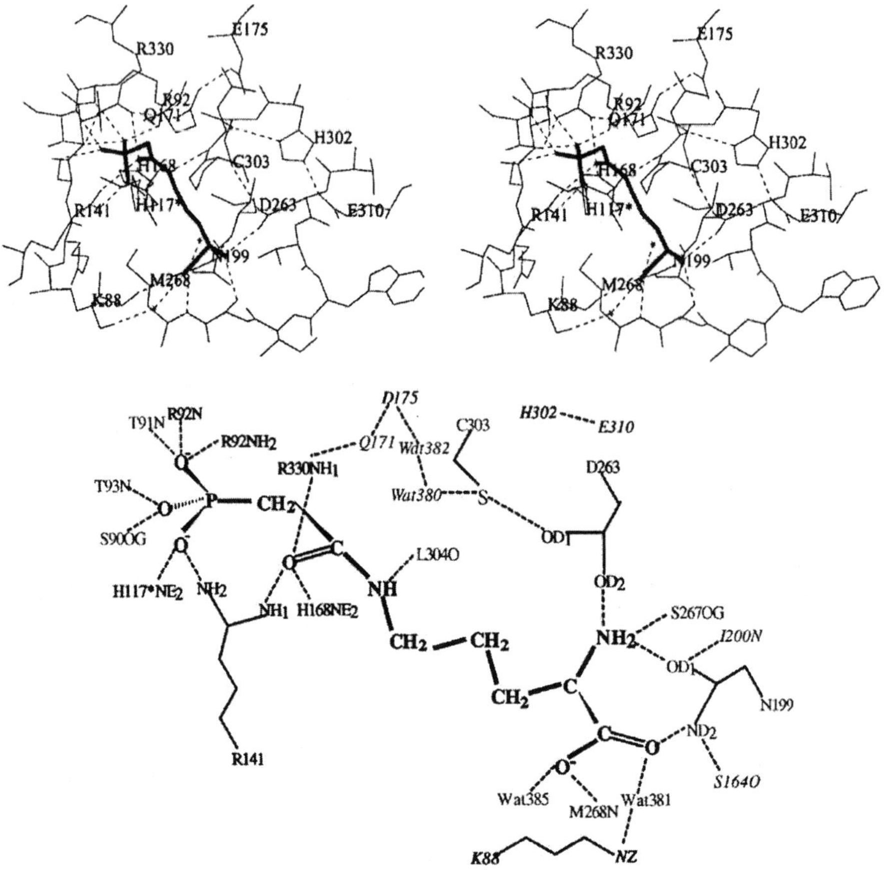

Figure 4-5. Stereo view (upper panel) and schematic (lower panel) showing the interaction of the bisubstrate analog PALO with active site residues[171]. PALO is shown in *bold*. The residue indicated with * is from an adjacent subunit. Numbering is for pOTC. (Copyright 1998. Reprinted with permission of the Amer Soc Biochem & Molec Biol.)

Table 4-1. Invariant, highly conserved and homologous conserved amino acids in the sequence of 33 OTCs[a]

Invariant		Almost invariant[b]	Homologous substitutions[c]	
56K	146T	32K (A_1)	8R-K (E_1)	106V-I
58S	149E	54F (Y_1)	11L-V-I (E_1)	116L-V (A_1 F_1 M_1)
60R	163G	*59T* (G_1 L_1 T_1 M_1)	20E-D (Q_2 T_1)	125V-I (M_1 K_1 C_1)
61T	*167N*	62R (S_1)	25L-V-I (E_1)	127V-I
73G	188P	87G (R_1 S_1)	48K-R (K_1 L_1 M_1)	141L-V-I (M_1)
90E	*231D*	133D (N_3)	50L-V-I (M_1)	228L-V-I
94D	*235S*	142A (T_2 C_1)	53I-V (M_2)	230T-S (G_2)
109R	237G	164D (H_1 E_1)	56E-D (V1 T1)	245R-K (V_2)
126P	270H	166N (Y_1)	64S-T (A_6)	252Y-F (W_1)
129N	*271C*	180G (D_2 P_1)	68G-A	254V-I-L
131L	*272L*	221A (G_3)	74G-A	262A-G (L_1)
136H	273P	233W(F3 C1)	78F-Y (V_2 I_1 D_1)	282E-D (A_2)
137P	296E	*236M* (L_2)	79L-V-I (F_2 M_1)	291V-L-I (A_5)
139Q	297N	268F (V_4)	86L-I (F_3 M_3)	292F-W
143D	*298R*	289S (N_3 G_1)	91S-T (P_5 E_1)	295A-G
			95S-T (F_2 V_3)	304A-G (S_4)
			98V-I (N_2)	305V-L-I (A_4)
			99L-V-I (M_4)	310L-V-I (A_1 F_1)

a. Data derived from Labedan B et al.[172] and supplement of 33 sequences of OTCs[173]. Numbering is from human OTC monomer cDNA as in Fig.4–1. Underlined are substrate or catalytic residues for hOTC as in Table 4–2.

b. All but 1–4 residues are identical; exceptions are in parentheses (i.e.,V_2).

c. Homologous or interchangeable amino acids are L-I-V, R-K, E-D, S-T, Y-F-W, G-A with 1–6 exceptions in parentheses.

The mature monomer of catabolic paeOTC contains 127 amino acids which are identical with those in analogous positions of hOTC (Fig. 4-1), amounting to 41% of the 308 amino acids which can be aligned in the two enzymes. Adding in the 33 homologous amino acid changes that result in minimal changes in function gives 160/308 amino acids (52%)that are similar. This remarkable similarity among hOTC and the two bacterial OTCs allowed us to use their X-ray structures to build our computer model of hOTC and find that it matched closely the 3D structure of hOTC found by Shi et al.[171]. When the two bacterial enzyme sequences are compared, 193 amino acids are identical and 31 similar for a 67.5% consonance between their 332 paired residues, including the extra loop-helix 9∝-loop area.

Murata and Schachman[175] performed a sequence alignment between *E.coli* ATC and OTC which showed 18 gaps, 32% identity and 56% similarity. The gaps and insertions in ATC correspond to surface turns, loops and coils. Houghton[176] compared the sequences of *E.coli* K-12 ATC and OTC and found 25–40% similarity depending on the insertions/deletions of the two monomers. If functional homologies of amino acids were taken into consideration the match was almost 50%. We find the match between ecoATC and ecoOTC in Fig. 4-1 amounts to 39%: 80 identical plus 33 similar amino acids in 292 pairs for ecoOTC and ecoATC, attesting to their shared evolutionary origins. The similarities among the ecoATC, paeOTC, ecoOTC and hOTC trimers justified the use of Lipscomb's ecoATC X-ray diffraction data as search probes in working out the 3-D structures of paeOTC[167], ecoOTC[168,169] and hOTC[171]. The residues which bind CP in ATC in Lipscomb's X-ray diffraction studies are those which are matched in OTC per many studies. However, different binding groups evolved for ornithine in the OTCs than the groups in ATC which bind aspartate, but they occupy similar positions in the C-terminal end of the monomer.

In Table 4-2 we have listed the residues in ecoATC shown by X-ray diffraction to bind substrates, substrate analogues or inhibitors like phosphate, pyrophosphate, N-(phosphonacetyl)-L-aspartate (PALA), phosphonoacetamide, malonate and succinate. Also shown in Table 4-2 are the groups in ecoOTC which serve similar functions in binding N-(phosphonacetyl)-L-ornithine (PALO)[169]. The structures of PALA and PALO are shown in Fig. 4-6, using the numbering system for the atoms used by the authors who did the X-ray structure studies[165,171]. We have listed the groups in hOTC which probably play the same roles (Table 4-2) based on the X-ray diffraction studies of hOTC structure and binding of PALO[171] and of CP and norvaline[177] and on functional studies in both enzymes, to be discussed subsequently.

Table 4-3 lists some of the essential residues in ATC which maintain tertiary or quaternary structure of its trimer and the analogous groups in the OTCs which probably play the same roles. When PALA binds to ATC, the polar and equatorial domains of the monomer move closer together, the polar domain binding the CP moiety of PALA and the equatorial domain binding the aspartate moiety. The so-called 80s loop (residues 79–85) from an adjacent catalytic chain and the so-called 240s loop (residues 229–239) in the same chain also move

Table 4-2. Substrate binding and catalytically active residues in *E.coli* aspartate transcarbamylase (ATC) compared with *E.coli* and human OTCs

Residues at equivalent positions in monomers of OTC[b]			
E.coli ATC	*E.coli* OTC	Human OTC	Presumed function in ATC or OTC[a]
—	K 53	K 56	H-bonds NZ via water 381 to OT2
S 52	S 55	S 58	H-bonds to O2P via OG
T 53	T 56	T 59	Backbone N H-bonds to O3P
R 54	R 57	R 60	H-bonds N,NE,NH1,NH2 to O3P; NE to OA of CP in OTC
T 55	T 58	T 61	H-bonds N and OG1 to O2P; and OG1 to O1 of CP in OTC
S 80	Q 82	H 85	From adjacent monomer S OG binds to O1P,O3P; in hOTC NE2 of H85 binds to O1P
K 84	K 86	?N 89	From adjacent monomer, K NZ binds to O1P of CP and in ATC O2,O5 of asp
R 105	R 106	R 109	Salt-bridge from NH1 to O1P; H-bonds NH2 to O1
H 134	H 133	H136	H-bonds NE2 to O1 of CP
Q 137	Q 136	Q 139	OE1 H-bonds to NH2 of CP
R 167	—	—	Binds NE to O4, NH2 to O5 of asp
—	N 167	N 167	H-bonds OD1 to α-amino of orn; ND2 H-bonds to OT2, helping domain closure
R 229	D 231	D 231	R binds NE to O3,NH2 to O2 of asp; OD2 of D forms salt-bridge with α-NH_3^+ of orn
Q 231	—	—	Q binds NE2 to O3 of asp
—	S 235	S 235	OG H-bonds to $\alpha-NH_3^+$ of orn
—	M 236	M 236	H-bonds N to OT1;backbone to δ-C of orn by hydrophobic bonds
P 266	—	—	H-bonds O to NH2 of CP
L 267	L 274	L 272	H-bonds O to NH_2 of CP and ? NH2 of orn
—	C 273	C 271	H-bonds S to δ-NH2 of orn; O to NH_2 of CP
—	R 319	R 298	H-bonds NH1 to O1 and NH_2 of CP

a. O1P, O2P, O3P are phosphate oxygens of CP; O1 is carbonyl O, OA is ester O of CP;OT1 is the carbonyl and OT2 is the carboxyl oxygen of ornithine. Water 383 O1 binds to OT2, water 387 O1 to OT1.

For ATC, O2 is carboxyl, O3 carbonyl on β-COOH; O5 is carbonyl, O4 carboxyl of α-COOH of asp.

b. Y165, D162, E233 and E272 are also essential for activity of ecoATC by site -directed mutagenesis.

into a position to interact with PALA. Regions 50–55, 80–85 and 225–245 contribute residues which interact with PALA [162]. Krause et al.[161] list 23 polar and non-polar interactions between subunits C1 and C2 when PALA is bound to ecoATC. We have listed at the bottom of Table 4-3 14 additional polar bonds linking C1 to C2 of ecoATC[165].

The X-ray diffraction three-dimensional structure of catabolic *Pseudomonas aeruginosa* OTC[167] monomer is very similar to that of ecoOTC without substrates bound[168]. When PALO binds to ecoOTC there is a marked change in positions 232–256 (the 240s loop) compared with that in non-PALO bound *Ps. aeruginosa* OTC, bringing the CP and ornithine binding domains together, and altering many

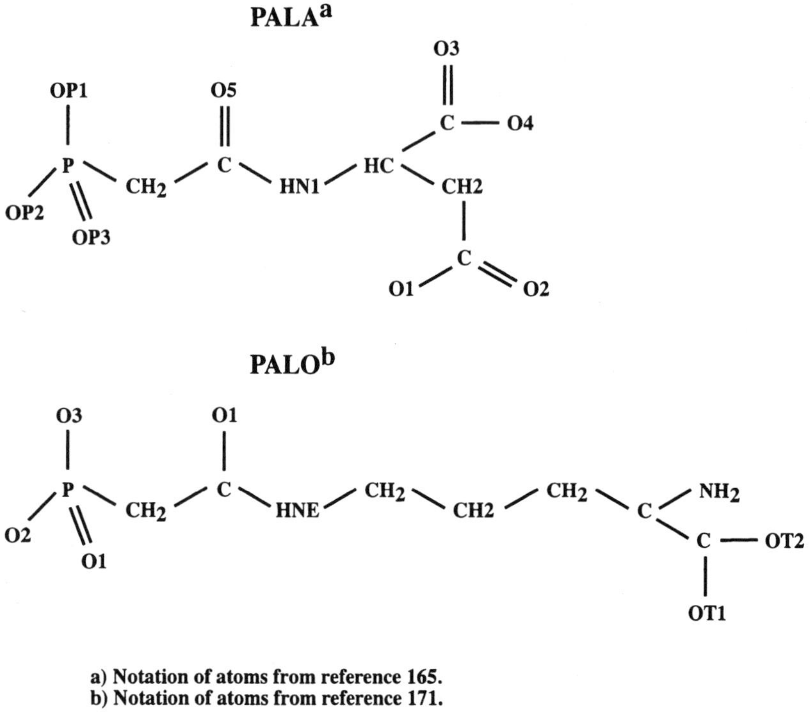

Figure 4-6. Structure and numbering of the atoms of N-(phosphonacetyl)-L-aspartate (PALA)[165] and N-(phosphonacetyl)-L-ornithine (PALO)[171], bisubstrate analogs of *E. coli* ATC and *E. coli* OTC, respectively.

Table 4-3. Essential structural residues in *E. coli* ATC compared with *E. coli* and human OTCs[a]

E.coli ATC[b]	*E.coli* OTC	Human OTC	Presumed functions
	within monomer		
K31-N291	K30-D314	K32-E296,N297	Links $\alpha1$, $\alpha11$ helices by polar bonds
K31-E147	K30-E146	K32-E149	Links $\alpha1$, $\alpha5$ helices by salt bridge
—	D163-H_2O-D231,V232	D164-H_2O-D231,T232	Positions SMG loop to bind orn
—	N167?-T129	N167-S132	Domain closure after PALO binds
—	M236-D54	M236-E55	When PALO binds,sulphur of 236 bonds to C-O of D or E
E233-R229	—	—	Positions R229 to bind β-COOH of asp
E239-K164	E 238-R165	E 239-N166?	When PALO binds, closes domain
—	R 246-D 163	R 245-D 164	No substrate bound, forms salt bridge,holds active site open
—	R 246-E 238	R 245-E 239 (C1-C3)	When PALO binds, closes domain
	between monomers		
E50-R167	E52-R165	—	Salt link from OE1 of E on C1 to NH1 of R on C3
E50-R234	E52-K240?	E55-R238	Salt link from C1 to C3
S52-N78	S55-Q82?	S58-D83	NH_2 of S H-bonds to O of D when substrates bind
E86-R54	E87-R57	E90-R60	Salt link of 80s loop of C1 to C3; in hOTC puts H85 near O1P
D90-R269	D91-H278?	D94-R274	Salt link from C1 to C3
V94-L267	V95-L274	V98-L272	Non-polar link from C1 to C3
—	—	R97-E282	Salt link from C1 to C3
D 100-R 65	—	—	Salt link from C1 to C3
—	M236-K53	—	NZ of K on C1 binds O of M on C3
—	N130-K53	—	OD1 of N on C1 binds N2 of K on C3

a. C1, C2, C3 are monomers of OTC trimer.

b. Additional polar bonds between C1-C2 of ATC: E37-K40, E37-H41, H41-R65, V43-R65, T53-S80, R54-S80, R54-Y98, R56-G72, H64-S69, H64-V70, R65-Y98, R65-D100, S74-N78, D75-N78, D75-N78, T97-G290[(121)].

noncovalent bonds as described by Ha et al.[169] showed that binding of PALO to ecoOTC causes a movement of the C_α positions of residues 236–256 of 6.5 angstroms toward the polar or CP end of the molecule. The C_α of G237 moves 9.5 angstroms. Moreover PALO binding to ecoOTC rotates the CP and ornithine binding domains by $\sim 9^0$, similar to the $\sim 6^0$ rotation of the aspartate domain of ecoATC when PALA binds[162].

Shi et al.[711] prepared hOTC crystals with PALO bound and then soaked the crystals with 100mM CP, changing the CP solution seven times, thereby displacing the reversibly bound PALO with CP and creating a binary complex of hOTC-CP. Crystallographic studies comparing the PALO-bound enzyme[171] and the CP-norvaline-bound enzyme[177] with the binary complex showed that binding of CP alone caused most of the domain reorientation which takes place with substrate binding. L-ornithine binding only brought the S235-M-G loop into the active site. This Shi study fits the report of Kuo and Seaton that binding of CP cracked single crystals of ecoOTC[217] and other reports that binding of CP alone induced the main UV absorbance changes seen in ecoOTC[216] and in hOTcc[712] in solution. In the binary hOTC structure three new water molecules filled the ornithine binding site, the SMG loop was less ordered and displaced from the active site and the adjacent loop, which brings H117 (H85 in the monomer)to bind CP in hOTC, was shifted 3.1Å judging by the C_α atom of V120[711]. The crystal structure of the binding site for CP is very similar to that in ecoATC, ecoOTC and hOTC in a transcarbamylase–like protein from the anaerobic bacterium *Bacteroides fragilis* which plays a role in arginine biosynthesis[712]. The second substrate is unknown and its binding site is quite different from those that bind aspartate or ornithine.

In addition to these X-ray diffraction structural studies, other functional methods have shown that a number of the same residues, as well as other groups in ATC are essential for substrate binding, catalysis or assembly of the trimeric holoenzyme. These other methods include kinetic studies, site-specific reagents and site-directed mutagenesis. The residues shown by *functional* studies to be essential for activity in ecoATC are histidine-134[178–180], tyrosine-165[181], threonine-55[182], arginines-54[183], -105[183]and -229[186], glutamine-137[183], serine-52[184], serine-80[184],aspartate-162[185], and glutamates-233 and

-272[186](Table 4-2). The residues of ATC shown in Table 4-3 that stabilize the tertiary or quaternary structure of the trimer have been confirmed or expanded upon by similar functional studies. The ATC amino acids shown essential for linkages between or within monomers by site-directed mutagenesis are serine-80[184], aspartate-100[187], arginine-269[187], glutamate-86[188], glutamate-50 linked to arginines-167 and -234[189], the residues of a C-terminal bridging helix 12[175] and tyrosine-165 of subunit C-1 H-bonded to glutamate-239 in subunit C-4 (a bond not present in OTCs)[190].

Functional studies of residues in ecoOTC and hOTC have helped support the roles suggested by structural and sequence studies of ATC (Table 4-2). *Histidines* have been implicated in OTC by a sigmoid-shaped curve derived from a plot of V_{max} vs. pH with a single ionization between 6.65 and 6.8[89]. This suggested that the ionization of an imidazolium cationic nitrogen of *histidine* at the active center was necessary in order for the catalytic step to take place. Grillo[147] carried out studies with diethylpyrocarbonate (DEPC) that reacts specifically at pH 6.l with histidine in proteins. He found that DEPC abolished activity and that CP protected against DEPC inhibition in beef liver OTC.

The studies of DeMars et al.[148] and Veres et al.[149] about the amino acid abnormality in the sparse fur mutant mouse showed that substitution of histidine by asparagine at residue 85 accounted for the shift in the pH optimum from 7.7 to 9 and a change from 6.7 to 8.7 in the pK_a of the group ionizing at the active center derived from V_{max} versus pH plots. In addition, the Km for ornithine was markedly increased but the activity at pH 9 with high ornithine concentrations was the same or greater than it was at pH 7.7. Immunologic cross-reacting material was increased. Both *E.coli* and human OTC contain histidines at residue 85 but they are not paired in a matched sequence (Fig. 4-2). Human, rat and mouse OTCs place this histidine in a sequence TQDIH(85)LGVNES, whereas ecoOTC and *Ps.aeruginosa* OTC have it in a sequence of PSGSQ(82)IGH(85)KES. Both are part of the 80s loop. The crystal structure of ecoOTC does not show binding of H-85 to PALO but does show H-bonding by Q-82 from an adjacent monomer to the phosphate oxygen-1 of PALO[169]. S-80 from an adjacent monomer in ecoATC H-bonds to this O-1P of PALA[159]. In Shi's X-ray structure of hOTC trimer[171], H-85 from an adjacent monomer binds to the O-1P of PALO (Table 4-2). Labedan's supplement showing the sequences of 33 OTCs has H at position 85 (numbering as in hOTC) in 6 species, Q in 22, N in 2, and G,R and A in one each[173].

Histidine-*134* in ATC binds to O1 of CP[178–180]and the analogous H-133 in ecoOTC H-bonds to the carbonyl-O of PALO[169]. Shi[171] found that H-136 of hOTC binds to the carbonyl-O of CP (Table 4-2). This bond is one of three that polarize the carbonyl-O of CP so a tetrahedral intermediate can form.

The second critical specific amino residue identified in human ornithine transcarbamylase is *cysteine* 271. This residue is found in an ornithine binding site in exon 9 which consists of HCLP (Fig. 4-1). This sequence is conserved throughout OTC's of 33 species[173]. Reichard[93] first showed in rat liver OTC that p-chloromercuribenzoate (PCMB) inhibits OTC and the inhibition could be partially reversed with SH reagents. He showed that carbamyl phosphate or ornithine protected 85% against this inhibition. Marshall and Cohen studied both the *Strept. faecalis* and the beef liver OTCs with various sulfhydryl reagents and showed that there was one SH-group per monomer which was essential for activity[191–195]. The S-thioethylamine derivative of OTC did not affect the dissociation constant for carbamyl phosphate and therefore it appeared from the experiments that the SH group was near the ornithine binding site. Marshall[195] showed that the cysteine involved was 52 amino acids from the C-terminus, which would place it at C271. Goldsmith[196] created a mutant of ecoOTC in which the C273 in this enzyme was replaced by alanine. The affinity for ornithine was 10-fold less in this mutant enzyme. A pKa of 6.2 was derived from the plot of log K_{cat}/K_m versus pH in the wild-type OTC. The pKa of 6.2 was lost in the C273A mutant so this pK_a was attributed to the ionization of C273. This mutant also showed impaired binding of norvaline, the inhibition constant being 0.5 mM compared to 0.054 mM in the wild type enzyme. Studies in rat liver OTC where C271S was created also suggested that cysteine H-bonded to ornithine[197]. Shi et al.[171]found in their structure of hOTC complexed with PALO that the sulfur of C271 H-bonded to OD1 of D263 whose OD2 H-bonds to the α-amino group of ornithine. In their subsequent paper where the crystal structure of hOTC was analyzed after complexing with CP and L-norvaline[177], the S-γ of C271 was close to the side chain of norvaline and its O atom formed an H-bond to the NP of CP. These findings were used by Shi et al. to propose that the sulfur of

C271 abstracts a proton from the δ-NH_3^+ of ornithine and serves as a general base in the catalytic reaction(171,177). Because this cysteine is found in 28 different OTCs(172,173), we can assign cysteine 271 as a critical but not the only binding residue for ornithine in human OTC. EcoATC has the sequence HPLP at the binding site for aspartate (Fig. 4-1), the proline-266 carbonyl-O H-bonding to the NH_2 of CP (Table 4-2).

Grillo(198) first suggested that essential *tyrosines* existed in beef OTC based on studies with various site-specific reagents(199) and concluded that tyrosine was at the active center near the ornithine binding site. In *ecoATC,* tyrosine-165 has been shown to be essential for catalysis, stabilizing the 160s loop(181), H-bonding from monomer C1 to E239 of the C4 monomer and linking the two trimers of ecoATC(190). This C1–C4 interaction does not occur in OTCs. In human OTC tyrosine is absent in this region. Thus we cannot find good evidence for a role for tyrosine in hOTC.

A critical part of the 160s loop is *asparagine*-167 in ecoOTC and hOTC whose OD1 H-bonds to the NH_2 of PALO(169,171), which is equivalent to the α-NH_3^+ of ornithine. The ND2 of N167 also H-bonds to one of the carboxylate oxygens of PALO (OT2), assisting in domain closure after ornithine binds. G163 (hOTC numbering) is invariant in 33 OTCs(173) and D164 is almost invariant (Table 4-1). We postulate that they also play a role in stabilizing the 160s loop and thus the internal structure of the monomer.

Marshall and Cohen(200) used pyridoxal-5-phosphate under restricted conditions to modify *lysine* groups of bovine or *Strep. faecalis* OTC. They were able to produce enzymes containing 1, 2 or 3 phospho-pyridoxal lysines per molecule, each on a different subunit. When one lysine was modified on each of the three subunits, bovine OTC bound carbamyl phosphate poorly but was active. Because ornithine cannot bind until carbamyl phosphate binds, the Km for ornithine was 600 times that in the normal enzyme. High concentrations of carbamyl phosphate displaced pyridoxal phosphate and a combination of carbamyl phosphate and norvaline protected against modification of the lysines. Marshall concluded that one lysine per monomer is near the binding site for carbamyl phosphate(200). Kalousek reported in an abstract(201) that pyridoxal-5-phosphate inactivated human and beef liver OTC by modifying one or more lysines on each monomer and that this reaction could be protected by carbamyl phosphate, confirming Marshall. Valentini et al.[202] modified OTC from the dolphin *Stenella* with pyridoxal phosphate, reduced the Schiff base with $NaBH_4$, and found that incorporation of one mol/mol of trimer was responsible for reducing catalytic activity to 20% of control. K_m values for the substrates were barely changed in the modified enzyme. The labeled peptides were isolated and sequenced; lysines-56 and -242 were modified but CP only protected lysine-56 from reaction with pyridoxal phosphate and maintained activity. Modification of lysine-242 did not seem to affect function. They postulated that the bulky pyridoxal phosphate on one monomer prevented the conformational changes in the other two monomers which occurs on CP binding[158]. The residual 20% of activity with normal K_ms they attributed to unmodified trimers.

In ecoOTC K-53 from an adjacent monomer binds to one carboxyl-O of PALO[169]via a water molecule. Labedan has found that the analogous lysine to lysine-56 in human OTC is found in all of 31 OTCs throughout widely differing phyla and species[173]. Shi et al.[171] found that the side chain amine group of K-56 from the adjacent monomer in hOTC binds to a water molecule which in turn binds to one carboxylate oxygen (OT2) of ornithine (Table 4-2).

Lysine-84 of ATC from an adjacent monomer was found to bind to CP in X-ray diffraction studies[161] and all 30 ATCs from different species have a K at position 84[173]. Lysine-86 of ecoOTC shows no direct bonds to PALO[169] but bonding via a water molecule is possible. Mutation of K-86 to Q-86 in ecoOTC lowered V_{max} to 0.8% of normal[175]. Human OTC contains asparagine-89 at this position. In only 12/33 OTCs is this residue a lysine as it is in ecoOTC, the alternatives being an asparagine in 10 species, a glycine in 7, an arginine in 3 and an aspartate in one. Shi et al. in hOTC suggest that N-89 could link to a phosphate oxygen (O1P) via a water molecule[171].

Marshall and Cohen[191] made an S-thioethylamine OTC by reacting it with cystamine. This *Strep. faecalis* OTC was then treated with high concentrations of iodoacetate which caused progressive loss of activity with time. The disulfide was then removed with an SH reagent and 14% of activity was restored. Amino acid analysis showed 1.3 carboxymethyl groups per monomer mostly existing as carboxymethyl-homocysteine, the product of *methionine* oxidation. The same results occurred with bovine OTC. Carbamyl phosphate bound to the carboxymethylated bovine enzyme but norvaline did not and the enzyme contained 1.46 carboxymethyl groups per monomer. Marshall raised the possibility that *methionine* is a hydrophobic binder of norvaline or ornithine. Methionine-236 is conserved in all of 27 of 29 OTCs[173] and may bind ornithine methylene groups as part of the highly conserved and postulated ornithine binding sequence SMG (residues 235–237). Genetic mutation of this methionine to threonine causes OTC deficiency with 5–7% residual activity (Table 5-2). Crystal structure of ecoOTC with bound PALO confirms that M-236 H-bonds its backbone NH to OT1 of PALO and presumably to ornithine (Table 4-2)[169]. M-236 in hOTC H-bonds its backbone NH to OT1[171,177].

Arginine was shown to be important as a result of an abstract by Kalousek and Rosenberg[201] wherein phenylglyoxal or 2,3-butanedione were added to the human and beef enzymes under conditions which are known to modify arginines. This led to 90% loss of activity with either reagent and the loss could be prevented by carbamyl phosphate or by ornithine. Site-directed mutagenesis of rat liver OTC, where arg 60 was replaced by leucine, resulted in a trimeric OTC with no activity[197]. A human mutation where arg 60 was replaced by gln resulted in neonatal death of the male infant Table 5-2). Finally, Marshall and Cohen[203] used butanedione and phenylglyoxal as site-specific reagents to identify one exceptionally reactive arginine involved in binding carbamyl phosphate in bovine OTC. One of the essential arginines found in ecoATC which binds carbamyl phosphate is arginine-54[161]. In ecoOTC an essential arginine per monomer near the binding site for CP was implied by phenylglyoxal studies[204] and mutagenesis of arginine-57 of ecoOTC to glycine-57 showed that the mutant bound CP less strongly, bound substrates randomly, did not undergo the conformational transitions which ornithine binding causes and also had a drastically decreased turnover rate due to weak transition state binding[205]. Further studies with a R57H mutant showed that when histidine is protonated the ordered binding sequence (CP, then ornithine) is maintained but the CP-induced fit isomerization is not seen[206]. If deprotonated, the histidine-57 enzyme binds substrates randomly. Lastly, kinetic studies suggested that a positive charge at residue 57 augments the turnover rate. Rynkiewicz [207] found that exogenous guanidines could chemically rescue the inactive mutant *E.coli* OTC R57G, raising the turnover to 10% of the wild-type enzyme. A partial positive charge was implicated in the rescue, but the kinetic mechanism remained random, not sequential, confirming Zambidis and Kuo[208] that the conformational change which occurs on CP binding is essential for sequential binding.

When arginine-106 of ecoOTC was replaced by glycine the enzyme developed new positive cooperativity for both substrates[209], a pattern only seen in wild-type OTC when Zn^{++} is bound to the ornithine site, inducing ornithine cooperativity[156–158]. Arginine-105 in ecoATC bonds by a salt-bridge to the O of phosphate (O1P)[161] and R-106 in ecoOTC does likewise[169]. In hOTC R109 forms polar bonds to O1P and to the carbonyl-O of PALO[171].

Grillo[198] oxidized *tryptophan* residues with N-bromo-succinimide (NBS) in bovine OTC and followed the loss of extinction at 280 nM to reflect loss of tryptophans. The enzyme was extremely sensitive to NBS, 0.036 mM causing 100% inhibition. SH groups were blocked with PCMB and activity was zero; OTC then was dialyzed against cysteine and activity was restored to 54% of normal. However, when incubation with PCMB was followed by NBS treatment, dialysis with cysteine to remove PCMB restored no activity. Carbamyl phosphate or ornithine protected against NBS. Grillo calculated from the change in absorption at 280 nM that one tryptophan per mole of enzyme was modified leading to 100% inhibition. He showed that there was no change in molecular weight, such as would occur if NBS led to disassembly of the trimer. Theoretically there should be one essential tryptophan in each monomer to get complete inhibition if the residues are essential for binding or catalysis. Shen[210] employed phase-modulation fluorescence measurements and site-directed mutagenesis to study *E.coli* OTC. He made single, double, and triple mutants of the four tryptophans, W -125, -192, -233 and -243 in each monomer, replacing them with isoleucine or phenylalanine. The conclusions from kinetic studies, NBS binding, fluorescence quantum yields and emission maxima were that tryptophans-233 and -243 are exposed and -125 and -192 are buried, and that tryptophan-233 and less so -243 are involved in ornithine binding and catalysis. Ha et al.[169] could not see such H-bonds from W-233 to the backbone carbons of PALO in their crystal structure of ecoOTC nor could Shi et al. in hOTC see binding of PALO to W233[171]. We have not included W233 in Table 4-2 but assume it plays a role in hOTC ornithine binding by hydrophobic interactions with the methylene groups of ornithine as part of the D231-T-W-I-S-M-G237 sequence. W233 is present in 25 of 29 different OTCs, with 3 Fs and one C being the alternatives[(173)].

The presumed catalytic mechanism of OTC is based on the structural and functional studies of ecoATC (vide supra) and the kinetic studies of Kuo et al.[(136)] in ecoOTC. Gouaux et al.[(211)] proposed in 1987 that in ecoATC CP and aspartate form a tetrahedral intermediate and that the breakdown of the intermediate was facilitated by an intramolecular proton transfer between the amino group of L-aspartate and a terminal phosphate oxygen of CP. This model was based on the X-ray diffraction studies of ATC binding various substrate analogues. The proton bound to the reacting amino group of aspartate interacts with a phosphate of CP to form a six-membered ring in a chair conformation by this model (Fig. 4-7A). The formation of a tetrahedral intermediate in ATC was supported by Stark[(212)] because of kinetic isotope effects with ^{14}C- and anhydride ^{18}O-labeled CP and in ecoOTC by Waldrop et al.[(213)] using ^{15}N-ornithine and ^{18}O-CP. Parmentier et al.[(214,215)] measured isotope effects of ^{13}C-CP and ^{15}N-aspartate in H_2O and D_2O for the catalytic subunit of ATC which were consistent with a nucleophilic attack on the carbonyl carbon of CP by the deprotonated α-amino group of aspartate (an addition) forming a tetrahedral intermediate. The enzyme polarizes the carbonyl group of CP by bonds from the C=O to R105, H134 and T55. An intramolecular proton transfer (an elimination) led to formation of the products. Kuo[(136)] proposed that the isomer of ornithine which binds and is reactive in ecoOTC is the $NH_2-(CH_2)_3-CH(NH_3^+)COO-$ zwitter ion form, based on his kinetic studies. This form amounts to only 17% of total ornithine at pH 8. We have adapted the mechanism shown by Parmentier[(214,215)] in ATC to OTC in Fig. 4-7C. The six-membered intermediate in a chair conformation (Fig. 4-7B) allows the phosphate group of CP, which needs to be protonated in order to become a better leaving group, to abstract a proton from the δ-amino group of ornithine. This intramolecular proton transfer leads to collapse of the intermediate and formation of products (Fig. 4-7C).

Kuo's kinetic analysis in ecoOTC of k_{cat} vs pH and k_{cat}/K_M (zwitter ion) vs pH[(136)] with both ornithine and the inhibitor norvaline which lacks a δ-amino group is *not* consistent with the binding of ornithine when the δ-amino is charged; rather the α-amino group must be charged to bind. Shi et al.[171] and Allewell et al.[713] suggested that the mechanism of OTC involves binding of a δ-NH_3^+ and an α-NH_2 form of ornithine, followed by depronation of the δ-NH3+ by the thiolate of C271. This mechanism is not consistent with Kuo's thorough kinetic study[100]. However binding of $NH_3^+-(CH_2)_3-CH(NH_3^+)COO^-$ which then would be converted to the δ-NH_2 form of ornithine by transfer of a proton to the S of C271 would be a preliminary step to the binding of the true δ-NH_2 substrate form of orni-

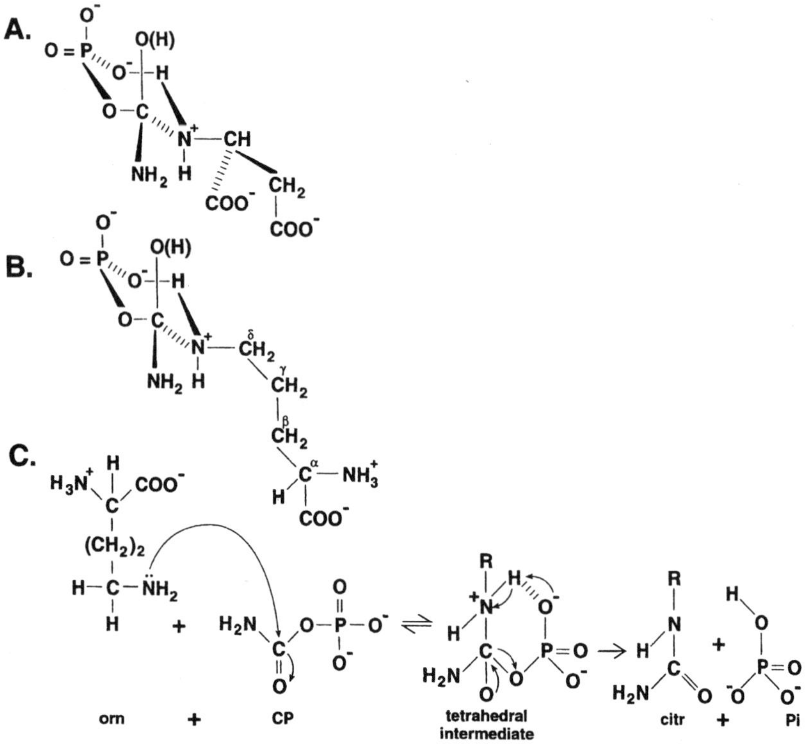

Figure 4-7. Catalytic mechanisms of ATC and OTC. A. Tetrahedral intermediate proposed for ATC[211]. B. Tetrahedral, six-membered intermediate proposed for OTC in a chair conformation. C. Proposed catalytic mechanism of OTC adapted from that of ATC[212–215]. (Copyright 1987. Reprinted with permission from Elsevier.)

thine and rationalize part of Shi's mechanism with that of Kuo[136]. Supporting functional evidence for Shi's mechanism is needed before we can accept that the form of ornithine that binds is $NH_3^+-(CH2)_3-CH(NH_2)COO^-$. We conclude that the chemical mechanism is that shown in Fig. 4-7C.

The kinetic mechanism for the OTC reaction we have chosen is adapted from that of Goldsmith and Kuo[196,206]:

$$E \underset{k_{-1}}{\overset{k_1(cp)}{\longleftrightarrow}} E\cdot cp \underset{k_{-iso}}{\overset{k_{iso}}{\longleftrightarrow}} E'\cdot cp \underset{k_{-2}}{\overset{k_2(orn)}{\longleftrightarrow}} E'\cdot cp\cdot orn \underset{k_{-3}}{\overset{k_3}{\longleftrightarrow}} [E'\cdot cp\cdot orn]^*$$

$$\underset{k_{-4}}{\overset{k_4}{\longleftrightarrow}} E\cdot P_i + cit \underset{k_{-5}}{\overset{k_5}{\longleftrightarrow}} E + P_i + cit$$

The enzyme binds carbamyl phosphate, as the dianion, in the cleft formed by helices ∝1,∝2 and ∝11 and by the 80s loop of an adjacent monomer. The enzyme undergoes an isomerization[206,216,217]due to binding to the groups in Fig. 4-6 with movements of arginine 60, of the 50s loop and of the 80s loop, forming a new isomer(E′)to which only the form of ornithine with an uncharged δ-amino group can bind. Thus the enzyme-substrate complex is formed by a sequential and ordered reaction[191,218]. The transition state $[E'\cdot cp\cdot orn]^*$ forms as shown in Fig. 4-7C and collapses, freeing citrulline first and inorganic phosphate (P_i)last. The rate-limiting reaction according to the kinetic analysis of Goldsmith and Kuo[206]is the rate of induced-fit isomerization (i.e., $k_{cat} \cong k_{iso}$). Older kinetic studies suggested that the slowest reaction was k_5, the dissociation of enzyme from phosphate.

Commentary. The linear amino acid sequences of OTCs from 33 species have been reported and all show amazing similarities especially in the substrate binding areas. Three dimensional structures of enzyme crystals by X-ray diffraction have been reported for the human, catabolic *Pseudomonas aeruginosa* and *E.coli* OTCs. Conserved and essential residues have been identified by structural and functional methods and compared with those of *E.coli* ATC. Crystallographic structures of the active centers with substrates bound and unbound show relative domain movements when carbamyl phosphate binds. When ornithine binds next, a flexible loop moves to close the active site. All studies are consistent with formation of a tetrahedral intermediate whose breakdown is facilitated by an intra molecular proton transfer between the δ-amino group of ornithine and a terminal phosphate of carbamyl phosphate. The likely kinetic mechanism includes an isomerization caused by CP binding and then a sequential, ordered reaction whereby ornithine binds to form a ternary transition complex which collapses, freeing citrulline first and lastly phosphate.

5

Molecular Pathology of OTC Deficiency

The diagnosis of OTC deficiency is made by assay of OTC activity in the liver of an affected proband, or by measurement of orotic acid +/- orotidine excretion in the urine after a protein or amino acid load or after a single dose of allopurinol in a proband or suspected carrier (see Chapter 8 for details). If prenatal diagnosis is desired, these approaches are less useful. OTC activity is not expressed in amniocytes and there are no unusual metabolites found in the amniotic fluid in OTC deficiency. Fetal liver biopsies are possible in order to assay urea cycle enzymes but are quite risky to the pregnancy[219–221]. Thus molecular genetic approaches have been developed to solve this problem.

Chromosomal deletions seen on cytogenetic analysis are rare[222–225,312]. However, amniocytes or chorionic villi biopsies can provide DNA for analysis. The DNA is extracted and digested with restriction endonucleases and the fragments are separated by electrophoresis and hybridized to full-length cDNA probes (Southern blots). Abnormal band patterns or restriction fragment length polymorphisms (RFLPs) may correlate with specific genetic defects. The same types of analysis can be performed on DNA from leukocytes, lymphoblasts, or fibroblasts in parents or siblings to compare with the fetus. However, after digestion of the DNA with restriction endonucleases used originally such as MspI, BamHI, HindIII or EcoRI only 14–17% of patients showed loss of an invariant band or a polymorphic band, appearance of a new band or loss of the entire gene[226–234]. The preferred method of prenatal diagnosis of OTC deficiency is now done by direct mutation analysis of chorionic villi, amniotic fluid cells or cultured amniocytes[235] by methods described below. Success is more likely if the mutation in an affected male family member or an obligate female carrier is identified before delivery. If no mutation in the fetus or family member can be found and there is more than one affected family member, then linkage analysis with markers close to the OTC locus on the X-chromosome can achieve 99% accuracy[234,235].

Prenatal diagnosis of OTC deficiency has been accomplished by using a single fetal nucleated erythrocyte isolated from maternal blood[236]. The entire genome of the one cell was amplified by primer extension preamplification. Next, sex and human leukocyte antigen-DQ genotype was determined from the DNA and then RFLP analysis and exon sequencing was carried out which showed that the fetus did not carry the exon 9 mutation of his brother.

Table 5-1 lists the first three RFLPs which proved useful in the diagnosis of presumed carriers or affected fetuses with OTC deficiency. MspI digestion yields two pairs of polymorphic bands and TaqI and BamHI give one pair each. The invariant bands are also listed for each in Table 5-1. Loss of one of these invariant bands and the appearance of a new band reflects the loss of a restriction site. For example, loss of the MspI band at 5.4-kb and appearance of a new band at 5.8-kb has been shown to result from loss of the site within exon 7 at nucleotides 675–678[237]. MspI restriction endonuclease digestion of seven OTC deficient male DNA isolates and probing with the PH0731 cDNA of Horwich[6] gave rise to six invariant bands and two polymorphic band

Table 5-1. Restriction endonucleases found useful in restriction fragment length polymorphism analysis of OTC deficiency

Restriction endonuclease	Sequence recognition	Polymorphic bands[a] (kilobase size)	Invariant bands[a]	Exon cleavage sites in cDNA
MspI	CCGG	6.6 / 6.2 5.1 / 4.4	17.5, 11.0, 5.4, 3.5, 2.0, 1.8	2, 7
TaqI	TCGA	3.7 / 3.6	4.6, 2.4	1, 3, 5, 9
Bam HI	GGATCC	18.0 / 5.2	20.0, 16.0, 9.6, 7.5, 6.6	

a. Bands obtained with pH0731 OTC cDNA [226].

sets at 6.6- and 6.2-kb and at 5.l- and 4.4-kb respectively (Table 5-1). Since each female gets one polymorphic band from each set, four haplotypes were possible: a 6.6- and 5.1-kb (A), a 6.6- and 4.4-kb (B), a 6.2- and 5.1-kb (C) and a 6.2- and 4.4-kb (D). The frequency of these haplotypes A, B, C and D among 35 females was A (45%), B (16%), C (28%) and D (11%)[(229)]. When the mother is suspected to be a carrier because she has given birth to an affected child then the RFLP pattern may allow assignment of the abnormal haplotype if she is heterozygous for the particular endonuclease site. For example, if an affected male has the 6.6- and 5.l-kb bands (haplotype A), and the carrier mother has the 6.6-, 6.2- and 5.l-kb bands (haplotype AC), then the A haplotype is associated with the defect.

Using full-length cDNAs the original polymorphic bands found useful were as follows: Bam HI 18- and 5.2-kb; MspI 6.6- and 6.2-kb and 5.l-and 4.4-kb; Taq I 3.7- and 3.6-kb using the Horwich-Rosenberg probe PH0731[(229)]. Pedigree RFLP studies indicated that 57% of heterozygous women result from new mutations in their father's sperm while there are rarely new mutations in ova (confidence interval 0–16%)[(238,239)], so the hemizygous, affected male almost always has an affected mother.

When RFLP analysis does not give information, then direct evaluation of the OTC gene is necessary. An algorithm adopted from Grompe[(311a)] for evaluation of families with OTC deficiency is shown in Figure 5-1. After OTC deficiency (OTCD) in a proband is detected by biochemical tests or by liver biopsy assay of OTC, a family history is taken from the core family which consists of the proband's immediate family and the mother's siblings and parents. If the history is negative, the core family is tested for carrier status by allopurinol tests with measurement of orotic acid and/or orotidine in the urine[(241)]. A positive allopurinol test, a clear result, leads to RFLP analysis. If there is a positive core family history, DNA is taken for a Southern blot and for RFLP studies with MspI, TaqI and some of the many other restriction endonucleases found useful in the OTC cDNA, some examples being MnII, MboII, AluI, RsaI and MseI (Table 5-2). Approximately 65% of the positive females are heterozygous for a known RFLP, have a deletion, or have the mutation detectable by TaqI digestion[(240)]. Families identified in this manner can use the RFLP for carrier or prenatal tests; many go on to DNA analysis of exons by PCR. In the mothers and core family members negative by history and ambiguous by allopurinol testing, plus those not heterozygous for an RFLP or lacking a deletion or the TaqI mutation, direct mutation detection is done by using a cDNA made from liver mRNA or by PCR expansion and sequence analysis of the 10 exons found in genomic DNA from leukocytes, fibroblasts or other somatic tissues. At least 90% of families will obtain an accurate diagnosis by using this algorithm (Fig. 5-1).

If there is only a singleton affected offspring with a known mutation and the mother tests negative, then the child is probably a new mutant(235,239). Individual exons can be amplified by PCR, followed by denaturing gel electrophoresis or single strand conformation polymorphism (SSCP) analysis and then direct sequencing of the exons. Denaturing gel electrophoresis is based on the effect of single base alterations on the melting behavior of a double-stranded DNA such that mutant and wild-type DNA can be separated in a denaturing gradient gel(242). SSCP(243)is described below. Another approach is to obtain liver specimens, make a cDNA from OTC mRNA, and form

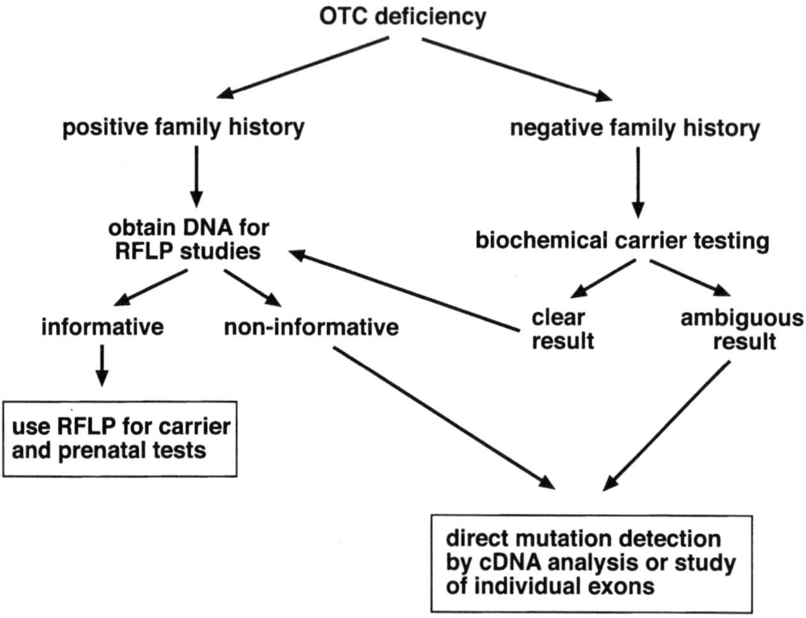

Figure 5-1. Algorithm for the diagnosis of OTC deficiency(311a). (Copyright 1990. Reprinted with permission from Elsevier.)

heteroduplexes between radiolabeled mutant DNA and wild-type DNA followed by chemical modification and cleavage at the site of the mutation(244).

One of the useful RFLPs is the Taq I endonuclease which recognizes the DNA sequence TCGA. Using a full-length human cDNA probe such as that described by Horwich(6), we should see the invariant bands (Table 5-1) which are produced on Southern blots after digestion with Taq I plus the polymorphic bands at 3.7- and 3.6-kb. The four cleavage sites for Taq I in the cDNA for pOTC are in exon 1, bases 66 to 69; exon 3, bases 273 to 276; exon 5, bases 420 to 423; and exon 9, bases 957 to 960 (Table 5-1). All of these sites contain a triplet ending in a T followed by a CGA codon for arginine. The CpG dinucleotide of TCGA is prone to mutation as discussed below. However TaqI RFLPs are the least common (3–5%) among the RFLPs useful in OTC deficiency. If there is a TaqI RFLP then the four exons which contain TaqI sites can be amplified by the polymerase chain reaction (PCR) using synthetic oligonucleotides which encompass the exon. The amplified products are digested with Taq I, the affected exon is thus identified and that exon is sequenced to find the nucleotide substitution.

The second restriction endonuclease that is informative is MspI which recognizes the sequence CCGG. It also contains the CpG prone to mutation. The MspI sites occur on exon 2, bases 146 to 149, and exon 7, bases 673 to 676 (Table 5-1). Digestion with other restriction endonucleases such as HindIII and EcoRI do not yield RFLPs but occasionally show new bands with diminution or loss of another band. Many new restriction endonucleases have been used to identify restriction sites on the OTC cDNA (Table 5-2). They are most useful after a sequence containing a mutation has been identified in a proband or female carrier. An endonuclease that recognizes that sequence can then be used as a short-cut in screening other family members. Overall, 80% of women at risk for having a son affected with OTC deficiency will be heterozygous for Msp1, Taq1 or BamH1. The percentages of control women who are heterozygous for an RFLP with MspI is 69%, for Bam HI is 29%, and for TaqI is 11%(231). If the OTC gene flanking markers 754 and LI.28 are also used, the percent of mothers with an affected child who have at least one RFLP that is informative rises to near 100%(234,235).

Loss of a restriction site does not necessarily mean that a functional defect in the OTC enzyme will occur. This is usually determined by the clinical history of the affected individuals. In one case, to prove that a missense mutation led to the OTC-deficient phenotype, a cDNA was synthesized containing the pOTC abnormality that occurred at base-pair 422, a C→T transition in the antisense strand that caused a sense-strand conversion of G→A, resulting in an arginine 141 (CGA) conversion to glutamine (CAA) at the TaqI site in exon 5[245]. When this mutant cDNA was transfected into COS-I cells there was no OTC activity whereas the normal cDNA produced activity in these cells, which ordinarily do not express OTC (Cos-1 is an immortalized African green monkey cell line). As shown in Table 4-2, pOTC arginine-141 (R109 on the mature monomer) is one of the critical binding sites of carbamyl phosphate.

In Table 5-2 we have compiled the reported and *sequenced* gene defects in OTC deficiency through December 2002 that have resulted in an affected male proband and those detected in females who may or may not be symptomatic[170,233,237,240,242,244–302,311,653–661,715–720]. We have been able to find 140 unique nucleotide changes in *males* among the 10 exons of the cDNA, causing nonsense, missense, frameshift or deletion mutations among the 1062 bases of the cDNA. All but one of these resulted in functional OTC deficiency and thus were not harmless polymorphisms. L111P was thought to be a harmless polymorphism, the P found in Horwich's cDNA sequence[6] and the L in Hata's sequence[8]. However, Grompe[244] found a male with the L111P variant who had tuberous sclerosis, "very mild" symptoms and no enzymic activity on a liver biopsy specimen. Oppliger-Leibundgut[282] sequenced 67 normals and found no CCT(P111)codons but all were CTT(L111). These normals were from Berne, Switzerland; Horwich's sequence with P111 came from a human cDNA library in New Haven, Conn. This problem could best be resolved by expressing the Horwich and Hata cDNAs in COS cells and studying their activities and stabilities *in vivo* and *in vitro*.

We also found 94 nucleotide changes in *females* who were symptomatic for OTC deficiency or who were proven to be obligate carriers after protein or allopurinol stress tests. The degree of OTC deficiency is of course determined by lyonization. Thirty three *intron* splice errors were also identified. The 140 male exon mutations occurred in 190 different families. The 94 female exon mutations occurred in 112 different families but 45 families were already in the

Table 5-2. Gene defects in OTC deficiency

Mutation number	Nucleotide No., pOTC cDNA	cDNA nucleotide change	pOTC amino acid codon change[b]	Exon	Restriction site lost or gained	OTC activity, % of normal[c]	Onset[d] N=neonatal L=late O=asympt.	Reference[e]
Males								
1	?1–1071	del exons 1–10		1–10.		0	N	661
2	? 1–297	del exons 1–3	del exons 1–3	1–3		0	N	285
3	53	CAC→del A	H 18 frameshift, early ter	1	AvaII	~0	N	661
4	67	CGA→TGA[a]	R 23 ter, leader sequence	1	TaqI, AsuII	~0	N	240
5	77	CGG→CAG[a]	R 26 Q, leader sequence	1		0	N	244
6	115	GGC→TGC	G 39 C	2	HaeIII, Sau961	?	N	297
7	118	CGT→TGT[a]	R 40 C	2	HaeIII, Cfr13I	3	L	267,718
8	119	CGT→CAT[a]	R 40 H	2	MaeIII, Tsp451	1.3–30; 16 families	L	264, 267, 270, 280, 281, 286, 655, 718
9	134	CTA→CCA	L 45 P	2		~0	N	244
10	135	del A, CTA	frameshift, 63 ter	2		0	N	297
11	140	AAC→ATC	N 47 I	2		~0	N	285
12	141+1	ins G, frame shift	K 54 ter	2	HindIII, AluI, CviJI	0	N	292, 298
13	148	GGA→AGA[a]	G 50 R	2	MspI, HpaII	?	L	285
14	154	GAA→AAA	E 52 K	2	MboI	?	N	302

Table 5-2 (continued)

Mutation number	Nucleotide No., pOTC cDNA	cDNA nucleotide change	pOTC amino acid codon change[b]	Exon	Restriction site lost or gained	OTC activity, % of normal[c]	Onset[d] N=neonatal L=late O=asympt.	Reference[e]
15	156	GAA→GAT	E 52 D	2	Tsp5091	4	L	302
16	158	ATT→AGT	I 53 S	2	MseI,TspEI	~0	N	661
17	163	TAT→GAT	Y 55 D	2		1.5,28	L	292
18	177	ATG→ACG	M 56 T	2		?	L	285
19	219–297	del exon 3		3		0	N	9
20	227	TTA→TCA	L 76 S	3	BsrDI	~0	N	655
21	231	TTG→TTT	L 77 F	3	CviRI,MwoI	?	L	302
22	236	GGG→GAG	G 79 E	3		0	N	254
23	240	AAG→AAT	K 80 N	3	EcoRI, Tsp 5091	?	L	656
24	247	GGC→CGC	G 83 R	3	Bsu361, Eco81I, DdeI	~0	N	285
25	248	GGC→GAC	G 83 D	3	CvnI	~0	N	653
26	259	GAG→AAG	E 87 K	3	MseI	0	N	251
27	264	AAA→AAT	K 88 N	3		3,18, 23 (3 fam's)	N/L	260, 281, 282
28	269	AGT→AAT	S 90 N	3	ScaI, RsaI	?	N	302
29	274	CGA→TGA[a]	R 92 ter	3	TaqI	0	N	240
30	275	CGA→CAA[a]	R 92 Q	3	TaqI	0	N	240
31	277	ACA→GCA	T 93 A	3	AvaI, XhoI	?	L	271
32	281	AGA→ACA	R 94 T	3		0.8	L	254
33	? 298–1062	del exons 4–10	del exons 4–10	4–10		0	N	285
34	305	GCA→GAA	A 102 E	4	EspI	0; 2 families	N	260, 285
35	327	del T, TGT	frameshift, 120 ter	4		?	N	297

Mutation number	Nucleotide No., pOTC cDNA	cDNA nucleotide change	pOTC amino acid codon change[b]	Exon	Restriction site lost or gained	OTC activity, % of normal[c]	Onset[d] N=neonatal L=late O=asympt.	Reference[e]
36	332	CTT→CCT	L 111P	4		0, ?unstable	L	244
37	350	CAT→CTT	H 117 L	4	MsII	?	L	263
38	350	CAT→CGT	H 117 R	4		18	L	292
39	374	ACG→ATG[a]	T 125 M	4	NlaII	<1	N	281
40	377	GAC→GGC	D 126 G	4	SduI, BseSI	0.9	N	265
41	386	CGT→CTT	R 129 L+donor splice error	4		6–20; 1 family	L	259
42	386	CGT→CAT[a]	R 129 H +donor splice error	4	MspI, HpaI	1.3–45; 6 families	L	265, 271, 281
43	?388–867	del exons 5–8		5–8		somatic mosaicism	L	223
44	?388–540	del exon 5		5		0; 2 families	N	655
45	396	TCT insT	frameshift,132 ter	5		~0	N	263
46	403	GCA→del G	A 135 frameshift, early ter	5	SphI, Nsp75241	~0	N	254
47	421	CGA→TGA[a]	R 141 ter	5	TaqI, XhoI	0; 4 families	N	240, 246, 256, 282
48	422	CGA→CAA[a]	R 141 Q	5	TaqI, XhoI	0; 5 families	N	240, 246, 252, 256
49	425	GTG→GAG	V 142 E	5			L	661
50	437	TCA→TGA	S 146 ter	5	BsrDI, ACRS	~0	N	655
51	460	GAA→TAA	E 154 ter	5		~0	N	244
52	476	ATT→ACT	I 159 T	5	Tsp5091	1.5	L	302
53	481	AAT→GAT	N 161 D	5	TaqI	~0	N	655
54	484	GGG→AGG	G 162 R	5		0	N	251

Table 5-2 (continued)

Mutation number	Nucleotide No., pOTC cDNA	cDNA nucleotide change	pOTC amino acid codon change[b]	Exon	Restriction site lost or gained	OTC activity, % of normal[c]	Onset[d] N=neonatal L=late O=asympt.	Reference[e]
55	491	TCA→TGA	S 164 ter	5	Hpy1881	0	N	258
56	493	GAT→TAT	D 165 Y	5		?	L	655
57	501	TAC→TAG	Y 167 ter	5	SfaNI, RsaI	0	N	292
58	501	TAC→TAA	Y 167 ter	5	MaeIII, RsaI	0	N	271
59	504	CAT→CAA	H 168 Q	5	BstF51, FokI	?	L	285
60	506	CCT→CTT	P 169 L	5	DdeI, ACRS	~0	N	655
61	516	ATC→ATG	I 172 M	5	EcoRII, DpnI, MvaI	0; 2 families	N	265, 268
62	524	GAT→GGT	D 175 G	5	BstEII	?	L	655
63	526	TAC→CAC	Y 176 H	5	BclI, MboI		N	661
64	527	TAC→TGC	Y 176 C	5		19	L	282
65	del 532–537	del ACGCTC	del T 178+L 179	5	EcoRII	0	N	292
66	533	ACG→ATG[a]	T 178 M	5	NlaIII	0; 2 families	N	267
67	534	dup TCAC	frameshift	5	BsaXI	0	N	281
68	540	CAG→CAC	Q 180 H+donor splice error	5	ScrFI, EcoRII, ApyI	~0, 7.1; 2 families	N, L	268, 292
69	542	GAA→GGA	E 181 G	6	BsmFI	~0	N	170
70	545	CAC→CTC	H 182 L	6		~0	N	268
71	548	TAT→TGT	Y 183 C	6	HpyCH4III, TspRI	0	N	253, 260

Mutation number	Nucleotide No., pOTC cDNA	cDNA nucleotide change	pOTC amino acid codon change[b]	Exon	Restriction site lost or gained	OTC activity, % of normal[c]	Onset[d] N=neonatal L=late O=asympt.	Reference[e]
72	562	GGT→CGT	G 188 R	6	HpyCH4III, TaiI	2	N	282
73	571	CTC→TTC	L 191 F	6	DdeI	6	L	660
74	576	AGC→AGG	S 192 R	6	PvuII, AluI	0	N	255
75	578	TGG→TAG	W 193 ter	6	PvuII, NspBII, MaeI	0	N	292
76	583	GGG→AGG[a]	G 195 R	6		0	N	302
77	586	GAT→TAT	D 196 Y	6	BstF5I, FokI	~0	N	170
78	587	GAT→GTT	D 196 V	6	BstF5I, FokI	3.7, 7	N	170, 255
79	596	AAT→AGT	N 199 S	6	HpyCH4III,BciVI		N	661
80	597–598	del TA	N 199 frameshift, early ter	6		0	N	264
81	602	CTG→CCG	L 201 P	6	FauI, BsgI	0	N	292
82	604	CAC→TAC	H 202 Y	6	TatI, RsaI	?	L	285
83	613	ATG→GTG	M 205 V	6	NlaIII	~0	N	655
84	614	ATG→ACG	M 205 T	6	Nlalll	?	N	296
85	620	ATG→AGG	M 206 R	6	MnII, MsII	?	N	285
86	621	AGC→AGA	S 207 R	6	CfoI, HhaI, HinPII	0	N	292
87	622	GCA→ACA[a]	A 208 T	6	ItaI, CfoI, HinPII	3, 4, 6; 3 families	L	284, 285, 287
88	626	GCG→GTG [a]	A 209 V	6	TseI, BtsI, BsoFI	0, 1; 2 families	N	281, 292
89	640	CAC→TAC	H 214 Y	6	RsaI	~0	N	279, 296
90	646	CAG→GAG	Q 216 E	6	MnII, SmII	~0	N	244

Table 5-2 (continued)

Mutation number	Nucleotide No., pOTC cDNA	cDNA nucleotide change	pOTC amino acid codon change[b]	Exon	Restriction site lost or gained	OTC activity, % of normal[c]	Onset[d] N=neonatal L=late O=asympt.	Reference[e]
91	646	insT, CAG→TCA	Q 216 S, frameshift, early ter	6	Eco571	0	N	263
92	658	CCA→GCA	P 220 A	6	BsII, MwoI	35	L	282
93	663	AAG→AAA	K 221 K, donor splice error	6		8	L	292
94	?664–1005	del exons 7–9	del exons 7–9	7—9	somatic mosaicism	?	L	311
95	?664–867	del exons 7–8	del exons 7–8	7—8		?	N	252
96	673	CCG→ACG	P 225 T	7	MspI, HpaI	?	L	264
97	674	CCG→CTG[a]	P 225 L	7	MspI, HpaI, AluI	0, 2; 2 families	N	237, 292
98	674	CCG→CGG	P 225 R	7	MspI, FauI	0	N	483
99	725	ACC→ATC	T 242 I	8	NlaIV, KpnI, RsaI, BanI	?	L	285
100	731	CTG→CAG	L 244 Q	8	AluI, CviJI	8	N	297
101	740	ACA→AAA	T 247 K	8		0	N	270
102	785	ACA→AAA	T 262 K	8		26	L	658
103	790	ACT→GCT	T 264 A	8	HgaI	22	L	255
104	791	ACT→ATT	T 264 I	8	HgaI	2	L	292
105	794	TGG→TTG	W 265 L	8		56, 59; 2 families	L	658
106	799	AGC→CGC	S 267 R	8	BciVI	?	L	302
107	803	ATG→ACG	M 268 T	8	NlaIII	5, 6.7; 2 families	L	255

Mutation number	Nucleotide No., pOTC cDNA	cDNA nucleotide change	pOTC amino acid codon change[b]	Exon	Restriction site lost or gained	OTC activity, % of normal[c]	Onset[d] N=neonatal L=late O=asympt.	Reference[e]
108	806	GGA→GAA	G 269 E	8	BsmFI	2	N	270
109	808	CAA→TAA	Q 270 ter	8	BsmFI	?	N	302
110	813–814	del AG>insC	frameshift, 288 ter	8	MnII	?0	N	300
111	816–818	del G, G, A	del E 272–273 in frame	8	MnII, BseRI	5	L	283
112	829	CGG→TGG[a]	R 277 W	8	ItaI, Fnu4HI	2–22; 8 families	L	242, 248, 255, 267
113	830	CGG→CAG[a]	R 277 Q	8	AluI, Bbv I, NlaIV	50–60; 3 families	L	264, 266
114	830	CGG→CTG	R 277 L	8	AciI, TseI		L	661
115	835	CAG→TAG	Q 279 ter	8	ApyI, MvaI	?	N	302
116	867	AAG→AAC	K 289 N	8	BsrGI, TatI	0	N	302
117	?868–1005	del exon 9	del exon 9	9		?	N	252
118	882	GCT→del T	A 294, frameshift, early ter	9	CfoI, HinPI, Fnu4HI	~0	N	257, 260
119	903	TTA→TTT	L 301 F	9		2.6, 3	L	519, 660, 717
120	904	CAC→TAC	H 302 Y	9	BtsI, TspRI	0	N	282
121	905	CAC→CGC	H 302 R	9	BsoFI	~0	N	655
122	906	CAC→CAG	H 302 Q	9	BtsI, HpyCH4III	?	L	285
123	907	TGC→CGC	C 303 R	9	AciI, BtsI	~0	N	297
124	907	TGC→GGC	C 303 G	9	CvinJI, BtsI	~0	N	661
125	912	TTG→TTT	L 304 F+ splice error	9	Cac81, MwoI	3.7–6; 3 families	L	9, 258, 273

Table 5-2 (continued)

Mutation number	Nucleotide No., pOTC cDNA	cDNA nucleotide change	pOTC amino acid codon change[b]	Exon	Restriction site lost or gained	OTC activity, % of normal[c]	Onset[d] N=neonatal L=late O=asympt.	Reference[e]
126	925–927	del GAA	del E 309, in frame deletion	9	MboII	2.4	L	264
127	928	GAA→TAA	E 310 ter	9	MboII	~0	N	257, 260
128	958	CGA→TGA[a]	R 320 ter	9	TaqI, Bcl I	0	N	292, 296
129	959	CGA→CTA	R 320 L	9	TaqI, DpnI, Sau3A	~0	N	240
130	962	TCA→TAA	S 321 ter	9	DpnI, MboI	~0	N	661
131	988	AGA→GGA	R 330 G	9		?	N	302
132	994	TGG→AGG	W 332 R	9		0	N	716
133	996	TGG→TGA	W 332 ter	9		6	N	265
134	1005	ATG→ATA	M 335 I	10	NlaIII	~0	N	661
135	1006	GCT→TCT	A 336 S	10	CviJI	?	L	170
136	1009	GTC→CTC	V337 L	10		6	L	292
137	1015	GTG→CTG	V 339 L	10	MwoI	~0	N	285
138	1028	ACA→AAA	T 343 K	10		14	L	285
139	1034	TAC→TGC	Y 345 C	10		?	N	170
140	1061	TTT→TGT	F 354 C	10	ApoI, FsiI, Tsp5091	1.8	L	285
Females								
1	1	ATG→GTG	M 1 V	1	SfaNI, MboII	40	L	291
2	3	ATG→ATA	M 1 I	1	SfaNI	34	L	660
3	67	CGA→TGA[a]	R 23 ter	1	TaqI, AsuII	16	L	292
4	77	CGG→CAG[a]	R 26 Q	1		?	N, L	291
5	94	CAA→TAA	Q 32 ter	2		17	L	291
6	106	CAG→TAG	Q 36 ter	2	Fnu4HI	?	L	655

Mutation number	Nucleotide No., pOTC cDNA	cDNA nucleotide change	pOTC amino acid codon change[b]	Exon	Restriction site lost or gained	OTC activity, % of normal[c]	Onset[d] N=neonatal L=late O=asympt.	Reference[e]
7	118	CGT→TGT[a]	R 40 C	2	HaeIII, Cfr13I	?	L	267
8	127	CTC→TTC	L 43 F	2		?	L	291
9	131	ACT→ATT	T 44 I	2		?	L	279
10	133	CTA→GTA	L 45 V	2	HpyCH4III, TspRI	?	L	170
11	140	AAC→ATC	N 47 I	2		?	L	302
12	143	TTT→TCT	F 48 S	2	BfaI	?	L	655
13	148	GGA→TGA	G 50 ter	2	DdeI, MspI, HpaII	4	L	251
14	154	GAA→TAA	E 52 ter	2	MboII	?	L	302
15	179	TCA→TTA	S 60 L	2		?; 2 families	L	268, 285
16	188	CTG→CCG	L 63 P	2	BglII, XhoII	?	L	291
17	205	CAG→TAG	Q 69 ter	2		?	L	301
18	238	AAG→GAG	K 80 E	3		?	L	657
19	243–245	del CTT	del L 82	3	Ddel, Eco8II, Bsu361	?	L	270
20	245	TTA→TGA	L 82 ter	3	BseMII, MnII	?	L	302
21	270	AGT→AGG	S 90 R	3	ScaI, RsaI	?	L	170
22	271	ACT→CT	T 91 del A	3	ScaI	?	L	655
23	274	CGA→TGA[a]	R 92 ter	3	TaqI	25; 3 families	N, L	281, 291, 715
24	275	CGA→CAA[a]	R 92 Q	3	TaqI	?	L	657
25	284	TTG→TCG	L 95 S	3	DpnI, MboI	?	L	302
26	299	GGC→GAC	G 100 D	4	AccI, EspI	?	L	291
27	316	GGA→AGA	G 106 R	4	MboII, MnII	?	L	302

Table 5-2 (continued)

Mutation number	Nucleotide No., pOTC cDNA	cDNA nucleotide change	pOTC amino acid codon change[b]	Exon	Restriction site lost or gained	OTC activity, % of normal[c]	Onset[d] N=neonatal L=late O=asympt.	Reference[e]
28	386	CGT→CTT	R 129 L+donor splice error	4		?	L	249
29	386	CGT→CAT[a]	R 129 H+donor splice error	4	MspI, HpaI	?, 3.5; 2 families	L	264, 271
30	388–393	ins TTA	V 130 ins L	5		?	L	661
31	416	TTG→TCG	L 139 S	5		?	L	285
32	421	CGA→TGA[a]	R 141 ter	5	TaqI, XhoI	4.5–5; 5 families	L	233, 246, 247, 256, 281
33	422	CGA→CCA	R 141 P	5	TaqI, XhoI	?	L	285
34	422	CGA→CAA[a]	R141 Q	5	TaqI, XhoI	?; 4 families	L	249, 256, 279, 296, 433
35	430	AAA→TAA	L 144 ter	5	MseI	?	L	270
36	443	TTG→TGG	L 148 W	5	BsmFI	?	L	302
37	444	TTG→TTC	L 148 F	5	TaqI	14	L	288
38	449–451	ACC→del C	T 150 frameshift, early ter	5	ApyI, EcoRII		L	661
39	455	GCT→GTT	A 152 V	5	CviJI, MseI	3.7	L	253, 294
40	463	GCA→TCA	A 155 S	5	HinfI, TfiI	?	L	661
41	476	ATT→ACT	I 159 T	5	Tsp5091	4.5	L	271
42	479	ATC→AGC	I 160 S	5	BsrDI		L	660
43	482	AAT→AGT	N 161 S	5		?, 19.4; 2 families	L	270, 719

Mutation number	Nucleotide No., pOTC cDNA	cDNA nucleotide change	pOTC amino acid codon change[b]	Exon	Restriction site lost or gained	OTC activity, % of normal[c]	Onset[d] N=neonatal L=late O=asympt.	Reference[e]
44	491	TCA→TGA	S 164 ter	5	Hpy1881	33	L	292
45	503	CAT→CGT	H 168 R	5	FokI, BstF5I	100, unstable	L	275
46	505	CCT→GCT	P 169 A	5	FokI, NlaIII	?	L	661
47	514	ATC→TTC	I 172 F	5	XhoII, DpnI	?	L	301
48	520	GCT→CCT	A 174 P	5	EcoRII, BspNI, ScrFI	?	L	261
49	524	GAT→GTT	D 175 V	5		?	L	285
50	533	ACG→ATG[a]	T 178 M	5	NlaIII	26	L	267
51	547	TAT→GAT	Y 183 D	6	SfcI	19	N	291
52	548	TAT→TGT	Y 183 C	6	HpyCH4III, TspRI	?	L	260
53	564	GGT→GTT	G 188 V	6		?	L	301
54	583	GGG→AGG[a]	G 195 R	6		0–6; 6 families	L	253, 264, 265, 291, 294, 655, 720
55	586	GAT→AT, delG	D 196 frameshift, early ter	6			L	660
56	589	GGG→AGG	G 197 R	6	FokI, BstF5I	?	L	301
57	590	GGG→GAG	G 197 E	6		?	L	170
58	594	AAC→AAA	N 198 K	6		?	L	299
59	605	CAC→CCC	H 202 P	6	BsgI, MnlI	100, unstable	L	295
60	608	TCC→TGC	S 203 C	6	TspRI, BtsI	?	L	263
61	618	ATG→ATC	M 206 I	6	MboI	5	L	660
62	622	GCA→ACA[a]	A 208 T	6	CfoI, HinP1I	58, 67; 1 family	L	287
63	626	GCG→GTG[a]	A 209 V	6	BsoFI	25	L	271, 299

Table 5-2 (continued)

Mutation number	Nucleotide No., pOTC cDNA	cDNA nucleotide change	pOTC amino acid codon change[b]	Exon	Restriction site lost or gained	OTC activity, % of normal[c]	Onset[d] N=neonatal L=late O=asympt.	Reference[e]
64	638	ATG→AAG	M 213 K	6	HpyCH4V, BsmI	?	L	291
65	640	CAC→TAC	H 214 Y	6	BsmI, RsaI	?; 1 family	L	279, 296
66	664	del G	del G 222,frameshift, ter	7		?	L	285, 293
67	674	CCG→CTG[a]	P 225 L	7	MspI, HpaI, AluI	?	L	291
68	717	GAG→GAA	E 239 E, donor splice error	7	MnlI	?	L	285
69	731–739	del 9 nt	del L244–248, frameshift	8	AluI	?	L	297
70	764	CAT→CCT	H 255 P	8	NlaIII, BpmI	?	N	170
71	787	GAC→AAC	D 263 N	8		?	L	285
72	788	GAC→GGC	D 263 G	8		?	L	170
73	790	ACT→GCT	T 264 A	8	HgaI	13	L	658
74	799	AGC→CGC	S 267 R	8	BciVI	21	L	292
75	829	CGG→TGG[a]	R 277 W	8	Fmn4HI	?	L	242
76	892–893	TGG→del TG	W298, frame shift	9	BslI, BsrI	?32	L	278
77	904	CAC→TAC	H 302 Y	9	TspRI	?	L	655
78	905	CAC→CTC	H 302 L	9	BtsI, TspRI	?	L	281
79	908	TGC→TAC	C 303 Y	9	BtsI, TspRI	?	L	285
80	914	CCC→CAC	P 305 H	9		46	L	660
81	928	GAA→TAA	E 310 ter	9	MboII	?	L	658
82	944	GTC→GGC	V 315 G	9	CviJI	?	L	302
83	944	GTC→GAC	V 315 D	9	BbsI, MboII	?	L	302

Mutation number	Nucleotide No., pOTC cDNA	cDNA nucleotide change	pOTC amino acid codon change[b]	Exon	Restriction site lost or gained	OTC activity, % of normal[c]	Onset[d] N=neonatal L=late O=asympt.	Reference[e]
84	947	TTT→TCT	F 316 S	9	BsmAI	?	L	302
85	953	TCT→TTT	S 318 F	9	BseRI	?	L	655
86	958	CGA→TGA[a]	R 320 ter	9	BclI, TaqI	?, ?; 2 families	L	279, 292, 296
87	976	GAG→AAG	E 326 K	9	MnII	?	L	299
88	988	AGA→GGA	R 330 G	9		?	L	285
89	988	delAGA, insT	R 330, frameshift, early ter	9	DdeI	18	L	660
90	1018	TCC→CCC	S 340 P	10	BanI, BsmFI	?	L	291
91	1022	CTG→CCG	L 341 P	10	AciI	8	L	660
92	1028	ACA→AAA	T 343 K	10		?	L	270
93	1033	TAC→GAC	Y 345 D	10	HinfI, PleI	3–16; 1 family	L	254
94	1042	CAG→TAG	Q 348 ter	10	MnII, AluI, BbvCI	?	N	291
			INTRON SPLICE ERRORS					
1		IVS1nt+1 g→t	donor splice site error		MseI	?, female	L	285
2		IVS1nt+4 a→c	splice error, low mRNA		TspRI, RsaI	0	N	277
3		IVS1nt+5 g→a	donor splice site error			0	N	285
4		IVS1nt+3–6	del aagt, donor splice error		HpHI	?, female	L	661
5		IVS2nt+1 g→a	donor site error, ?skip E2			0; 2 families	N	282, 292

Table 5-2 (continued)

Mutation number	Nucleotide No., pOTC cDNA	cDNA nucleotide change	pOTC amino acid codon change[b]	Exon	Restriction site lost or gained	OTC activity, % of normal[c]	Onset[d] N=neonatal L=late O=asympt.	Reference[e]
6		IVS2nt-1 g→a	acceptor site lost		PsiI, SspI	0	N	254
7		IVS3nt+1 g→a	16 aa's added to exon 3			0.6	N	272
8		IVS3nt+1 delgtaag	donor splice site		HpyCH4III	?, female	L	285
9		IVS4nt-2 a→t	acceptor site, 4 aa's lost			0	N	250
10		IVS4nt-2 a→c	acceptor site error		TspRI	?	N	302
11		IVS4nt-2 a→g	acceptor site error		TspRI, BtgI	?	N	302
12		IVS4nt-12 g→a	acceptor site error			?, female	L	258
13		IVS5nt+1 g→c	donor site error		EcoRII, AluI	0	N	282
14		IVS5nt+2 t→c	exon 5 deleted		CviJI	0	N	273
15		IVS5nt+3 t→tt	donor splice site error			?, female	L	270
16		IVS5nt-2 a→g	acceptor site error		XcmI, ACRS	~0	N	655
17		IVS6nt+1 g→t	donor site error			50, female; 2 fam's	L	270, 282
18		IVS6nt+1 g→a	donor site error			?, female	L	285

Mutation number	Nucleotide No., pOTC cDNA	cDNA nucleotide change	pOTC amino acid codon change[b]	Exon	Restriction site lost or gained	OTC activity, % of normal[c]	Onset[d] N=neonatal L=late O=asympt.	Reference[e]
19		IVS6nt-1g→a	acceptor site error		EcoNI, MseI	?, female	N, L	660, 661
20		IVS6nt+2 t→c	donor site error			?, female	L	285
21		IVS6nt+3 del g	donor site error			?, female	L	302
22		IVS6nt+3 t→tt	donor site error		BsII	?, female	L	270
23		IVS7nt+1 g→a	donor site error		MnlI	~0	N	655
24		IVS7nt+1 g→t	donor site error		MnlI	?, female	L	661
25		IVS7nt+2 t→c	donor site, exon 7 lost	7	BglI, SphI	0	N	250, 302
26		IVS7nt+3 a→g	" " "	7	BsiHKAI	0	N	250
27		IVS7nt7–22del16nt	splice error		NlaIII, NspI	0–1.5	N	297
28		IVS7nt-2 a→g	acceptor site error		BfaI	?, female	L	299
29		IVS8nt+1 g→t	?E8 skipping	8	MaeIII, RsaI	0	N	282
30		IVS8nt+1 g→a	" "	8	MboII, RsaI	25, female	L	258,302
31		IVS9nt+1 g→t	donor splice error		MseI	0	N	285

Table 5-2 (continued)

Mutation number	Nucleotide No., pOTC cDNA	cDNA nucleotide change	pOTC amino acid codon change[b]	Exon	Restriction site lost or gained	OTC activity, % of normal[c]	Onset[d] N=neonatal L=late O=asympt.	Reference[e]
32		IVS9nt+2t→c	donor splice site error		Cac8I	~0	N	661
33		IVS9nt-3c→g	acceptor site error		BspHI	3	L	660

Note: All reports through 2002.
a. CpG doublet mutations; antisense C→T causes sense strand G→A.
b. To obtain OTC codon, subtract 32 (length of leader peptide) from pOTC codon.
c. % of mean or upper normal as reported by author.
d. Onset=neonatal if 1–28 days of age; late if >28 days.
e. Cases not included unless each was sequenced.

male series for a total of *257 unique families* with exon-localized OTC deficiency. Among these 140 males there were 15 nonsense mutants with early termination, 103 missense changes, 10 frameshifts and 11 deletions. Nonsense, missense, frameshift or deletions among males occur in each exon as follows: exon 1(5), exon 2(13), exon 3(14), exon 4(10), exon 5(26), exon 6(25), exon 7(5), exon 8(18), exon 9(17) and exon 10(7).

Onset of symptoms of OTC deficiency were noted in 90 male mutants by 1–30 days, defined as early or neonatal onset and most of these had no or very low liver OTC activity. Fifty had the onset of clinical symptoms beyond 30 days defined as late onset. Mutations #27 and #68 in Table 5-2 occurred in families with both a neonatal and a late onset course. We combined the 257 unique families with male or female exon mutations and calculated the frequency of unique (family) defects in each exon by dividing the number of affected families by the number of nucleotides per exon: exon **1**(7/78=0.090), **2**(41/138=0.297), **3**(21/81=0.259), **4**(18/90=0.200), **5**(52/153=0.340), **6**(38/123=0.309), **7**(8/54=0.148), **8**(33/150=0.220), **9**(30/138=0.217) and **10**(11/57=0.193). The *rate* at which mutations occur in families is highest in exons 5 and 6 and lowest in exons 1 and 7. The exon 5 rate may reflect an ascertainment bias because one of the four TaqI sites is found here. Codon 141(CGA) for arginine, a TaqI site, in exon 5 accounted for 9 male and 10 female mutations alone. The intron errors are fairly evenly distributed (Table 5-2).

Oppliger-Leibundgut[267] sequenced the entire coding region in 30 cases of OTC deficiency and found that 11 (37%) of errors were at CpG dinucleotides, which only total 18 in the coding region of the OTC monomer. In Table 2 the commonest mutation is that of a CG→TG on either strand of the DNA, marked with an *a*. Such errors are thought to occur because 50% of cytosines in these doublets are methylated and spontaneous or enzymatic oxidative deamination of the methyl-cytosine results in a thymine. When this base change occurs on the sense strand, a CGA for example is converted to TGA, which is a termination or a stop codon (male mutation 4, Table 5-2). There are 10 male and 9 female sites involved by sense-strand CpG mutations involving 24 male-derived and 16 female-derived families in the cDNA for OTC (Table 5-2). When a CpG change occurs on the antisense strand, for example TCG to TTG, it requires a balancing conversion on the sense strand of a G to A causing a conversion of a

CGA (arginine) to CAA (glutamine) (male mutation 30, Table 5-2). These anti-sense mutations are also marked with an *a*. There are 9 male and 6 female sites involved by such antisense-strand mutations (Table 5-2) involving 37 male-derived and 16 female-derived families. Thus 93 out of 257 unique families (36%) developed mutations at CpG doublets on the sense and antisense strands, clear evidence that these are "hot-spots" for mutations. CpG doublets that contain a methylated cytosine are "hot spots" for mutations not only in OTC but in other proteins(170,303). MspI sites (CCGG recognition sequence) also contain this CpG "hot spot" sequence where C-to-T transitions occur. Both MspI and all 4 TaqI sites (TCGA) in the pOTC cDNA have been mutated in Table 5-2. The frequency of these CpG mutations may be partly spurious because these RFLP sites are easier to find and to sequence using exon specific oligonucleotides.

Out of 18 CpG sites in the cDNA of pOTC, 15 had been reported as mutated by 2002. The 3 which had not been reported are T127 (ACG), F211(TTC)-G212(GCA) and A254(GCG). Five CpG mutations do not involve the expected C→T or G→A change. These are two males R129L (CGT→CTT) and R320L (CGA→CTA and three females G50ter (C-GGA→C-TGA), R129L (CGT→CTT) and R141P (CGA→CCA).

Some base changes cause no functional effects and appear to be polymorphisms (Table 5-3). Nine of these occur in exons and seven in introns. Plante and Tuchman(304) found that four of these polymorphisms (K46R, Q270R, IVS3-8nt, IVS4-7nt) yielded a range in frequency from 0.125 to 0.68 when the sequences of 45–76 suspected OTC deficiency patients were determined. The calculated proportion of hererozygosity in females in this population (2pq) ranged from 0.219 to 0.5. This allowed a calculation that 87% of females should be heterozygous for at least one of these 4 polymorphisms. These 4 sites can be used to do linkage analysis in the roughly 22% of patients with enzymatic OTC deficiency whose PCR sequencing does not detect a harmful mutation in the 10 exons or exon/intron borders(304).

The acceptor site mutations found among the intron splice errors of Table 5-2 occur on the 3′-end of the intron close to the splice site and interfere with the rule that an *ag* must signal the end of the intron. Errors also involve the donor site rule where *gt* must signal the beginning of an intron. In Table 5-2 mutation number 9 under intron errors(250) resulted from an *a* to *t* mutation at this *ag* acceptor site leading

Table 5-3. Polymorphisms detected in the OTC gene that have no functional significance

Exon Polymorphisms

Change in pOTC codon	OTC codon	Nucleotide Change	Exon	Reference
K 46 R	14	AAA->AGA	2	8,244
F 101 L	69	TTT->TTA	4	8, 270
L 111 P	79	CTT->CCT	4	8, 244[a]
Y 143 Y	111	TAT->TAC	5	304
W 193 C	161	TGG->TGT	6	8, 254
I 194 F	162	ATC->TTC	6	8
G 195 G	163	GGG->GGA	6	170
Q 270 R	238	CAA->CGA	8	8, 254
T 333 A	301	ACA->GCA	9	660

Intron Polymorphisms

Type of Change	Nucleotide Change	Intron	
IVS2 -54nt	ins a(ac->aac)	2	256
IVS3 -8nt	a->t	3	270
IVS3 -17nt	ins t(ct->ctt)	3	256
IVS4 +8nt	a->g	4	270
IVS4 -7nt	a->g	4	256
IVS6 +3nt	ins t(gta->gtta)	6	270
IVS9 -12nt	g->t	9	256

a. *P 111 has no activity in liver *in vitro* per Grompe[240] but Horwich[6] found it in a normal human DNA library; it may be unstable *in vitro*.

to a loss of 12 bases and four amino acids as the splicing apparatus searched down the exon for a new splice site. In mutation number 25, a *t* was converted to a *c* in the *gt* donor site and the whole of exon 7 preceding intron 7 was not spliced. Speculation about the mechanism of this error is contained in some detail in the paper by Carstens[250]. Carstens postulates elimination of an exon when a donor site from the previous intron joins to the next intron, having to do with the mechanism of the splicing complex as it runs from 5′- to 3′- on the pre-mRNA.

The advantage of finding the mutation is that it is not necessary to have an informative family history confirmed by RFLPs. The defect found in the proband can be sought by sequencing the gene in the mother to determine if this is a new mutation or if the mother is a carrier. It is possible to create 17 to 20-length oligonucleotides with DNA synthesizers and use these as 5′- and 3′- probes for each of the 10 OTC exons. Focusing on one involved exon is greatly helped by finding a restriction endonuclease cleavage site. When such a site is not available then denaturing gradient gel electrophoresis has been found useful(242). PCR is done to amplify the exons and then the hybridization is done in a progressive denaturing gel with a linear gradient. The pairs of bands that are heteroduplexes have a nonmatching area and "melt" or separate in acrylamide containing 40% formamide and 7 M urea. In a female carrier, two heteroduplex bands are usually found, a mutant and a normal band. The abnormal fragment then can be sequenced. Another method for detection of mismatched bases is that of Cotton et al.(305) In this method osmium tetroxide and piperidine cleave T or C mismatches in heteroduplex DNAs; hyroxylamine followed by piperidine cleaves C mismatches.

However, the most common method used for detection of single nucleotide substitutions or small deletions or insertions in DNA is by single-strand conformation polymorphism analysis (SSCP)(243). After isolation of high molecular weight DNA from leukocytes or liver, all ten exons of the OTC gene are amplified by the polymerase chain reaction (PCR) using primers for each exon and adjacent introns. Some of the primers do not detect splicing donor or acceptor sites and miss a few bases at the ends of exons. Thus SSCP may miss ~10% of the coding region mutations(261). Moreover, PCR amplifies only the normal allele in heterozygote females when there is a large deletion.

In SSCP, PCR products are denatured to separate strands and electrophoresed on a 12% polyacrylamide gel containing 10% glycerol, using a Tris-borate-EDTA buffer at pH 8, and then stained. DNA bands that migrate abnormally are purified using a 2% agarose gel, excised, purified again and sequenced. The mobility shifts caused by single base changes are not always detectable. The sensitivity of SSCP for OTC mutations is close to but not 100%(261). As can be judged by the above description SSCP is a long and complex procedure.

We have attempted to explain how the mutations in Table 5-2 cause the defects in function and structure that occur. Table 5-4 lists the presumed mechanisms why mutations cause neonatal onset of OTC deficiency in males. Missense mutations were evaluated comparing the role of the normal amino acid with the expected change in function or structure by the mutant amino acid, using empirical predictions of secondary protein structure in key articles and a book on protein structure[306–310]. We inspected the site of each missense mutation on our computer model of the 3-dimensional structure of OTC, looking for possible changes in the structure of α-helices, of β-strands and of loops by substitution of the mutant amino acids, or of unwanted side-chain interactions with adjacent chains. This proved to give little insight into why functions might suffer. If we had made separate computer models with the mutant amino acid inserted, we might have recognized deleterious changes in secondary or tertiary structure. Our conclusion was that neither empirical rules of secondary structure nor 3-dimensional structures derived from X-ray crystallography had reached the point where they could predict reliably whether a missense mutation would cause a functional defect or not in the OTC enzyme, with the exception of a proline insertion breaking a helix. However, we had the experiments of nature to guide us: the evolutionary history of 33 OTCs allowed us to compare which amino acids are accepted functionally at any site when aligned properly with each other according to the publication of Labedan et al.[172] and their supplement[173] which they made available to us.

We constructed Tables 5-4, 5-5 and 5-6 giving possible mechanisms for mutations that cause male neonatal onset, male late onset or female heterozygote cases respectively of OTC deficiency. We used Table 4-1 which lists the invariant, almost invariant and homologous substitutions in 33 OTCs found by Labedan[172,173] and Tables 4-2 and 4-3 which list the substrate-binding sites and essential structural sites to explain why the missense mutations in Table 5-1 might not be acceptable and result in functional defects. For active site and essential structural residues this comparison works quite well. For other missense mutations the evolutionary approach raises many questions. For example, in neonatal male 61 (I172M), why is M accepted in 5 other OTCs but not in human, mouse or rat OTC? Codon 172 is an isoleucine in the human, mouse and rat, but an M in catabolic *Ps.aerug.* or an L in *E.coli* OTC although the X-ray crystallographic structures of human, *E.coli* and *Ps.aerug.* OTCs are almost identical in this polar region of the molecule. The strict conservation of most residues among the human, rat and mouse sequences contrasts with the acceptance of non-homologous residues in sequences of bacteria, plants, fungi, etc. In only 21 codons does human OTC differ from mouse OTC and none are the invariant, almost invariant or homologous substitution sites listed in Table 4-1. The evolutionary tree constructed from these 33 OTCs[172] emphasizes the almost identical se-

Table 5-4. Possible mechanisms for missense mutations that cause neonatal onset of OTC deficiency in males

Mutation number[a]	Mutation, pOTC codon	Enzyme activity[b]	Possible mechanism or comparisons among other OTCs[c]
Active site missense mutations			
27	K 88 N	3	invariant; ?N does not H-bond via water 381 to OT2 of orn
28	S 90 N	~0	invariant; ?N does not H-bond to O2P, D115 on adjacent monomer
30	R 92 Q	~0	invariant; no salt-bridge from C1 to E122 of C3
48	R 141 Q	~0	invariant; loss of salt-bridge to O2P
123	C 303 R	~0	invariant; ?loss of H-bond to δ-NH_2 of orn and NH_2 of CP
124	C 303 G	~0	invariant; ?loss of H-bond to δ-NH2 of orn and NH2 of CP
131	R 330 G	~0	invariant; loss of H-bond to O1 and NH2 of CP
Structural missense mutations—within monomer			
69	E 181 G	~0	No polar bond to K 64, linking α1 and α5 helices
77	D 196 Y	~0	No H-bonds via water molecule to D 263, T264
78	D 196 V	3.7, 7	Same as above; D263 positions S267-M-G loop to bind orn
Structural missense mutations—between monomers			
26	E 87 K	~0	No salt-bridge from C1 to R 270 on C3
40	D 126 G	~0	No salt-bridge from C1 to R 306 on C3
Other missense mutations			
5	R 26 Q	0	Loss of + charge in leader sequence
6	G 39 C	~0	In other OTCs accepts L, P, A, V, I, N, Q, K, H in this position
9	L 45 P	~0	P breaks α1a helix;accepts I, M, F, V in other OTCs
11	N 47 I	~0	Accepts D, E, G, V, H; why not I?
14	E 52 K	~0	Accepts T, D, Q; why not K in loop before α1 helix?
16	I 53 S	~0	Accepts S, F, L, V, Q, M in other OTCs; why not in hOTC?
20	L 76 S	~0	Accepts S, K, P, A, V, I?
22	G 79 E	~0	Accepts N, R, H, S; why not E?
24	G 83 R	~0	Accepts A, S, V; why not R in β2?
25	G 83 D	~0	In middle of β2; accepts A, V, S, C but not D
34	A 102 E	0	Accepts S, T, V, Y, I, K, F; why not E at C1 position of α2 helix?

Mutation number[a]	Mutation, pOTC codon	Enzyme activity[b]	Possible mechanism or comparisons among other OTCs[c]
39	T 125 M	~0	Accepts K, Y, A, R, S, E, G, Q; M not preferred in N1 position of helix
53	N 161 D	~0	N invariant in $\beta 5$
54	G 162 R	0	Accepts A, S not R at end of $\beta 5$ strand
60	P 169 L	0	P invariant at HPxQ binding for CP
61	I 172 M	0	Accepts M in 5 OTCs plus A, V, L?
63	Y 176 H	~0	Accepts L, I, F, A, M, V; why not H?
66	T 178 M	0	T invariant in $\alpha 5$ bridge
70	H 182 L	~0	Accepts L in one OTC plus S, T, A, K, V, Y, I, N, E, R, C, F in others?
71	Y 183 C	0	Accepts C in one OTC plus F, A, L, S?
72	G 188 R	2	Accepts K, N, A, Q, E, D but not R in other OTCs
74	S 192 R	0	Accepts A, V, T but not R?
76	G 195 R	0	Invariant in loop from $\beta 6$ to $\alpha 6$ helix
79	N 199 S	~0	N invariant ; binds to ornithine
81	L 201 P	0	P is β-breaker at start of $\alpha 6$ helix; accepts I, V, A, C, T, S, G
83	M 205 V	~0	Accepts C, A, L, I but not V?
84	M 205 T	~0	Accepts C, A, L, I but not T?
85	M 206 R	~0	Accepts I, L, Q, E, Y but not R?
86	S 207 R	0	Accepts A, L, V, G, E, T, I but not R?
88	A 209 V	~0	Accepts L, T, R, P, I, M, and V in one OTC?
89	H 214 Y	~0	Accepts S, D, K, E, Q, N, T but not Y in $\beta 7$?
90	Q 216 E	~0	E is β-breaker in $\beta 7$ strand; accepts H, S, A, V, R, Y, T
97	P 225 L	0, 2	Accepts M, V, F, I, K, E but not L?
98	P 225 R	0	In loop before $\alpha 7$; accepts I, V, F, M, K, E but not R?
100	L 244 Q	8	Accepts F, I, V, E, S, T, K, R but nor Q?
101	T 247 K	0	Accepts V, S, F, L at end of $\beta 8$
108	G 269 E	2	G invariant at end of S-M-G 237 loop
116	K 289 N	0	Accepts N in one OTC plus E, D, A, R
120	H 302 Y	~0	H invariant, part of H-C-L-P which binds CP, orn
121	H 302 R	~0	H invariant, part of H-C-L-P which binds CP, orn
129	R 320 L	~0	Accepts K, H, N, Y, Q, A, M but not L?
132	W 332 R	0	Accepts H but not positive charge on R
134	M 335 I	~0	Accepts K, not I?
137	V 339 L	~0	Accepts L in 5 other OTCs plus D, E, K, I, V??
139	Y 345 C	~0	Accepts R, K, S in 7 OTCs but not C?

a. Numbers correspond to those in Table 5.2.

b. % of normal liver activity. When activity not measured, ~0 means close to zero is assumed.

c. Comparisons of alignments of 33 OTCs [172,173].

quences among the 3 mammals. The variation in residues accepted at many codons in other OTCs but not in human OTC while the other OTCs maintain the same catalytic function remains a mystery.

We did not list on Table 5-4 the nonsense mutations or frameshifts which give a termination codon, result in a truncated monomer, and are always neonatal in onset in males (Table 5-2). W332ter still gave 6% of normal activity even with deletion of the last 22 amino acids!

Seven neonatal-type missense mutations affect active site residues, and five affect essential structural groups (Table 5-4). Among the remaining 45 neonatal missense errors, only five can be predicted to have a deleterious effect and that is when a proline addition or removal changes the direction of a loop or breaks an α-helix (L45P, P169L, L201P, P225L, P225R). The rest of these missense errors have mechanistic explanations that are speculative. Site-directed mutagenesis of each of these remaining errors in human OTC followed by transfection into COS-1 cells could give a clue to functional enzymatic defects[(294)]. X-ray diffraction of mutant crystalline enzymes may help explain some defects but is time-consuming[(721)].

The hypothesis has been proposed that missense, neonatal mutations that cause essentially zero OTC activity are located within the interior of the monomer near the active site or substrate binding residues or are part of the monomeric or trimeric bridging loops, while late onset defects with measurable activity are remote from the active site or located on the surface of the monomer or trimer[(268,661)]. Figure 2 in Tuchman et al.[(661)] attempts to illustrate this hypothesis but it is apparent by inspection that the neonatal mutations are sometimes on the surface and late-onset mutations are often near the active center and substrate-binding regions. Female mutations occur in both central and peripheral regions of the monomer, as we would expect from Table 5-2.

Table 5-5 collates the missense mutations that cause *late onset* OTC deficiency in males. Seven mutations affect active site residues and six affect essential structural elements. The 31 other late-onset mutations include three with losses or gains of a proline (L111P, P220A, P225T). Only 13 mutations lie on the surface of the monomer or trimer and 9 lie inside the monomer, according to our inspection of Figure 4-4 and Figure 2 in Tuchman et al. As in the neonatal group deleterious mutations in human OTC are acceptable variants in other species of OTC.

Table 5-6 contains 51 missense mutations in females for which we

Table 5-5. Possible mechanisms for missense mutations that cause late onset of OTC deficiency in males

Mutation number[a]	Mutation, pOTC codon	Enzyme activity[b]	Possible mechanism or comparisons among other OTCs[c]
Active site missense mutations			
31	T 93 A	?	Invariant;loss of H-bond to NH1 of R109
37	H 117 L	?	?Loss of H-bond to O1P but accepts Q, G, A, N, R in other OTCs
37	H 117 R	18	R accepted in one OTC?
59	H 168 Q	?	H invariant
106	S 267 R	?	S invariant in SMG loop; binds orn
107	M 268 T	5, 6.7	M almost invariant in SMG loop; accepts L in 2 OTCs
125	L 304 F	3.7–6	L invariant in HCLP loop which binds CP, orn
Structural missense mutations—within monomer			
103	T 264 A	9	OH of T H-bonds H_2O to D196; accepts V, I not A?
104	T 264 I	2	Accepts I in one OTC?
112	R 277 W	2 to 22	W breaks salt-bridge to D196 when no substrates are bound
113	R 277 Q	50–60	Q breaks salt-bridge to E271, which closes domain when substrates bind; accepts K, V but not Q or W?
Structural missense mutations—between monomers			
41	R 129 L	6 to 20	Salt-bridge from C1 to E314 on C3 broken
42	R 129 H	1.3–2.1	Same as # 41 but accepts K, V, Q, A?
Other missense mutations			
7	R 40 C	3	R on convex side of trimer, ? binding C-terminus by salt-bridge
8	R 40 H	1.3–30	Same as # 7; accepts K, and also E in one OTC
13	G 50 R	?	Accepts A, D, I, P, K, E, Q, T, and R in 2 OTCs?
15	E 52 D	4	Third residue of $\alpha 1$; accepts 2 Ds and Q, T in other OTCs
14	Y 55 D	1.5; 28	Accepts I, K, S, T, E, A, G, H, and D in one OTC?
17	M 56 T	?	Accepts L, I, V, Y; prefers non-polar groups
21	L 77 F	?	Third residue before $\beta 4$; accepts F in 3 OTCs; also M, H?

Table 5-5 (continued)

Mutation number[a]	Mutation, pOTC codon	Enzyme activity[b]	Possible mechanism or comparisons among other OTCs[c]
32	R 94 T	0.8	R almost invariant at end of STRTR motif;accepts one S, not T?
36	L 111 P	0 in vitro	Polymorphism; P in Horwich[(6)], L in Hata[(8)];accepts F, M, V, I
49	V 142 E	?	Accepts L, T, G; ?E binds to R109 so it can't bind CP
52	I 159 T	1.5	N-terminal of $\beta 5$; accepts V, L but not T?
56	D 165 Y	?	D is almost invariant on loop to H136; accepts 3 Ns
62	D 175 G	?	D is invariant, binds to Q171 and by water molecules to C303
64	Y 176 C	?	Accepts L, I, F, A, M, V, Y but not C?
68	Q 180 H	0, 7	Donor splice error;accepts I, K, Y, S, W, L, E, V, A, R; why not H?
73	L 191 F	6	Accepts F, I, M, V, Y; why not F in hOTC?
82	H 202 Y	?	Accepts N, F, T, R and one Y?
87	A 208 T	4, 6	Accepts C, N, S, G, E, F, A but not T?
92	P 220 A	?	P invariant after $\beta 7$strand
96	P 225 T	?	Accepts M, V, F, I, L, K, E but not T?
99	T 242 I	?	Accepts A, S, T, G, K, E, D but not I?
102	T 262 K	26	Accepts S, G; ?K binds to, interferes with D263
105	W 265 L	56, 59	Almost invariable;accepts F, C; pH mutant on loop to bind orn
106	S 267 R	?	S invariant in SMG loop; binds orn
114	R 277 L	?	R almost invariable; salt bridge to D164; accepts K but why V?
119	L 301 F	2.6, 3	Accepts L, many M; why not F?
122	H 302 Q	?	H invariant in HCLP motiv binding orn
135	A 336 S	?	Accepts G and 4 Ss in other OTCs?
136	V 337 L	6	Accepts A, I, and 5 Ls in other OTCs?
138	T 343 K	14	Accepts V, E, A, S, N, G in 16 OTCs; why not K?
140	F 354 C	?	C-Terminus; F binds hydrophobically to $\alpha 11$ helix; C does not?

a. Number corresponds to that in Table 5-2.
b. % of normal liver activity.
c. Comparisons of alignments of 33 OTCs[(172,173)].

have tried to explain the resulting defects. Eight novel mutations involve loss or gain of a proline which helps explain the loss of function in the mutant allele. As in the male mutants, hOTC does not accept the same or homologous substitutions which are tolerated in other OTCs. Some subtle differences between hOTC and ecoOTC or paeOTC, for example, not seen on X-ray studies of crystalline enzymes must account for these functional defects of hOTC in the mitochondria. Tuchman has reported that mutations found in symptomatic female probands have usually not been associated with late-onset disease in males, but rather with severe defects causing neonatal disease(722). In our review of Table 5-2 we found 9 females with low liver OTC activities whose mutations were identical to those of neonatal onset probands but contrary to Tuchman's report we found four mutations in 11 families with symptomatic females whose mutations were identical to those of *late-onset* males. They were R129H (6 families), I159T (1 family), A208T (3 families) and T264A (1 family).

Both R277W and R277Q mutants cause late-onset disease. When cDNAs of these mutants were transfected into *E.coli,* overexpressed, purified and studied kinetically(289), both showed poor ornithine binding, loss of substrate inhibition, an alkaline shift in pH optimum and impaired thermal stability. An explanation for these changes may be that R277 and D196 form an ionic bond when no substrate is bound and hold the active site open; R277 and E271 then form a salt-bridge when substrate binds and close the binding domain (Table 4-3). R277Q protein forms two crystal forms, one wild-type cubic and one bipyramidal form, whereas R277W protein forms only the abnormal bipyramidal crystals(721). This finding fits with the fact that the liver OTC activities of R277W patients are lower than those with the R277Q defect (Table 5-2).

The common late onset disease mutation R40H on pOTC (R8 on the mature monomer) has defied explanation as to why this change from a strong basic side chain (R) to a zwitter ion side chain (H) causes *in vitro* activities ranging from 1.3 to 30% of normal. This position contains an R in 22 other OTCs, a K in 6 OTCs and an E in one(173). Mutation number 5 with the R40C change had 3% of normal OTC activity. Thus H or C are not tolerated at codon 40 in OTCs. When the R40H pOTC was cloned and expressed in *E.coli,* then purified and studied kinetically, it did not differ from the wild-type OTC *in vitro*(290). Nishiyori(286) found that the liver OTC mRNA levels

Table 5-6. Possible mechanisms for effects of missense mutations in female heterozygotes with OTC deficiency

Mutation number[a]	Mutation, pOTC codon	Possible mechanism or comparisons among other OTCs[b]
1	M 1 V	M is initiation codon for OTC
2	M 1 I	Loss of initiation codon
8	L 43 F	Accepts I, V, E; why not F?
9	T 44 I	Accepts S, H, D, C, A, R, K, all polar
10	L 45 V	N-terminus of α1a; accepts only one V but I, M, F, E in other OTCs
12	F 48 S	Accepts Y, F, L, H; why not S?
15	S 60 L	Accepts L in one OTC, plus A, G
16	L 63 P	?P breaks α2 helix; accepts F, I, H, V, A, C
18	K 80 E	Accepts Q, M, L and 4Rs; E inserts a negative charge
21	S 90 R	Invariant;S H-Bonds to O2P in CP binding site
25	L 95 S	Accepts I, V, T, M, F, C; why not S?
26	G 100 D	Accepts only G and A
27	G 106 R	Accepts only G and A
31	L 139 S	Accepts F, M, A, V, E, Q; why not S?
33	R 141 P	Invariant; loss of salt-bridge to O2P of CP
36	L 148 W	Accepts V, I, M, F, A, all nonpolar
37	L 148 F	Fourth residue in α4; accepts one F, 17Vs, 3 Is, one M, A in 33 OTCs
39	A 152 V	Accepts V in 3 OTCs, plus C
40	A 155 S	Accepts S, G, P; why not S in hOTC? Is this a polymorphism?
42	I 160 S	Accepts V, Y, C, W, F; why not S?
43	N 161 S	N invariant; its function is unknown
45	H 168 R	Invariant; H-bonds to O1 of CP
46	P 169 A	P invariant in HPxQ binding site for CP
47	I 172 F	Accepts A, L, V, M; why not F?
48	A 174 P	?P breaks α5 helix; accepts C, T
49	D 175 V	Invariant; ? function of D in α5 helix
51	Y 183 D	Accepts F, L, A, S, C; negative charge of D a problem?
53	G 188 V	Accepts K, N, A, Q, E, D; why not V?
56	G 197 R	Accepts A, V, I, T; R positive charge a problem?
57	G 197 E	Same as # 56; E negative charge a problem?
58	N 198 K	Almost invariant;Y found in one OTC
59	H 202 P	?P a breaker of α6 helix; accepts N, F, T, R, Y
60	S 203 C	Accepts D, A; why not C?
61	M 206 I	Accepts I, L, Q, E, Y; OTC=5%; why is I not functional in hOTC?
64	M 213 K	Accepts I, V, F, L, A, C, Y; K a breaker of β7 strand?
70	H 255 P	?P changes direction of α8 loop; accepts 10 others

Mutation number[a]	Mutation, pOTC codon	Possible mechanism or comparisons among other OTCs[b]
71	D 263 N	Invariant; D forms salt-bridge to α-amino of orn
72	D 263 G	Same as #71
73	T 264 A	Accepts V, I; why not A in S235MG loop?
78	H 302 L	Invariant; part of HCLP motif binding CP and orn
79	C 303 Y	Invariant; C binds δ-amino of orn
80	P 305 H	P invariant in HCLP loop binding orn
82	V 315 G	?G a breaker of α10 helix;accepts A, I, L, T, S, E
83	V 315 D	Same as #82; why accept E but not D?
84	F 316 S	Accepts V, A, I, L, M in α10 helix, all non-polar
85	S 318 F	Accepts C, E, N, H, G; why not F?
87	E 326 K	Accepts Q or E but not K
88	R 330 G	R invariant, binds CP
90	S 340 P	?P a breaker of α11 helix;accepts I, A, G, H, T, F, L
91	L 341 P	Accepts F, V, T, A but P breaks α-11 helix
93	Y 345 D	Accepts R, K, S but not D at start of C-terminal loop

a. Numbers correspond to those in Table 5-2, female section.
b. Comparisons of alignments of 33 OTCs[(172, 173)].

were normal in an R40H patient whose activity was only 5% of normal. When the mutant mRNA was transfected into COS-1 cells, it translated as well as normal transfected OTC mRNA and was equally stable. But the activity of the R40H enzyme in these cells was only 28% of that in cells transfected with normal OTC. Five cycles of freeze/thaw did not affect normal OTC in COS-1 cells but caused the R40H enzyme to decrease to 17% of its initial activity. K_ms for both substrates and pH optima were the same as controls. These workers concluded that the R40H enzyme protein is unstable in the mitochondrion for unknown reasons. Mavinakere et al.[(654)] found that the R40H pOTC was imported normally into rat liver mitochondria. They showed that CHO cells transfected with the R40H cDNA made pOTC and imported part into mitochondria but degraded the rest of pOTC in the cytoplasm. R40 forms a salt bridge with E52 which may be essential in the structure of pOTC; H40, however, may be open to attack by proteases[(654)]. Morizono[(290)] suggested that the R40H mutation may change normal cleavage of the 32 amino acid leader peptide leaving an unstable, longer monomer.

Commentary. Prenatal or female carrier diagnosis of OTC deficiency has evolved to PCR amplification of the 10 exons followed by single strand conformation polymorphism (SSCP) analysis and sequencing of abnormal exons and intron borders. Restriction fragment length polymorphisms (RFLPs) are most useful when DNA sequence errors are known and restriction endonucleases are available which include the mutation. Our finding that 93 of the 257 unique family mutations (36%) occurred at the 18 CpG sites on OTC confirms that these are "hot spots" for mutations. When nonsense mutations cause a termination codon and a truncated monomer or when missense mutations occur at substrate-binding sites or essential structural sites it is obvious why severe disease resulted. However, other missense mutations that cause amino acid changes that result in severe disease of humans are often tolerated in reviewing sequences of 32 OTCs of other species. Explaining how certain mutations cause neonatal, severe disease and others cause late-onset, mild disease is not possible in many cases of human OTC deficiency.

6

Animal Models of OTC Deficiency and Their Gene Therapy

Sparse-Fur Mutant Mouse

In 1976 Demars[148] reported that a sparse fur mouse whose phenotype was due to an X-chromosomal defect was prone to develop bladder stones which on analysis turned out to be orotic acid. Because orotic aciduria occurs in human OTC deficiency, OTC activities of the livers of these sparse fur (spf) mice were assayed and it was found that they indeed had an OTC deficiency. The male spf mouse is smaller than its litter mates, has no fur and wrinkled skin at birth which persists from five to seven days and then disappears to a variable extent as the males develop some hair and grow somewhat less than normal. The OTC activity in this spf/Y mouse liver had an activity of 22% of the control Oak Ridge 22A or C57BL mouse liver enzyme at the usual pH optimum of 7.6 to 8[148](Table 6-1). However, male spf mice showed an increasing OTC activity as the pH of the medium was increased, the curve being sigmoidal in shape and reaching a maximum of 200% of the controls at pH 10. The midpoint of the sigmoid curve was 8.8. The normal mouse liver has an activity 80% of the spf/Y maximum at pH 10. The ornithine concentrations at half maximal velocity (K_m) were approximately 0.2 mM in the normal and 0.6 mM in the spf livers at pH 8[148]. When the K_m for carbamyl phosphate, leucine inhibition and heat stability were tested, normal and spf males were the same. At the mitochondrial pH of 7.4 the combination of

Table 6-1. Characteristics of the sparse-fur (spf) mouse model of OTC deficiency

Characteristics		spf		Wild-type	References
pOTC cDNA change		AAC (nt353–355)		CAC (nt353–355)	149
pOTC amino acid change		N 117		H 117	149
pOTC cDNA length		1062 bp		1062 bp	149
pOTC mRNA length		1650 bp		1650 bp	149
pOTC monomer size		40 kD		40 kD	316
OTC monomer		36–37 kD		36–37 kD	313, 316
V_{max} liver (μmol/hr/mgP), (pH 7.7, CP10mM, orn 0.7–10mM,	spf/Y	21+/-4	X/Y	157+/-42	315
fed 24% protein mouse chow)	spf/X	84+/-33	X/X	150+/-16	315
OTC activity(μmol/hr/mg enzyme)	spf/Y	24,000(pH 9)	X/Y	18,000(pH 8)	313
K_m liver(mM)					
CP(pH 7.7, orn 5mM, CP 0.7–10mM)	spf/Y	0.09+/-0.03	X/Y	0.95+/-0.3	315
	spf/X	0.45+/-0.28	X/X	0.99+/-0.3	315
Orn(pH 7.7, CP10mM, orn 0.7–10mM)	spf/Y	2.09+/-0.67	X/Y	1.67+/-0.29	315
	spf/X	1.40+/-0.4	X/X	1.65+/-0.2	315
CP(pH 8, orn 5mM, CP 0.04–5mM)	spf/Y	0.09	X/Y	0.15	313
Orn(pH 8, CP 5mM, orn 0.04–5mM)	spf/Y	2	X/Y	0.4	313
pH optimum	spf/Y	10		7.6–8	148, 313
Ratio, activity pH 9.5/8.0	spf/Y	6.2		0.89	313
Cross-reacting material (% control)	spf/Y	120, 150			313, 316

Characteristics		spf		Wild-type	References
Translatable mRNA (% control)	spf/Y	58			316
Synthesis rate, liver slices (% control)	spf/Y	62			316
pOTC mRNA (% control)	spf/Y	70, 74	X/Y	100	318
Orotic acid excretion, urinary (μmol/mg creatinine), 24% protein diet	spf/Y	18.4+/-5.1	X/Y	0.72+/-0.23	315
	spf/X	7.4+/-5.6	X/X	0.54+/-0.27	315
Plasma NH_3 (μg/dl), 24% protein diet	spf/Y	24–60	X/Y		317
	spf/X	16.0+/-0.97SE	X/X	14.7+/-0.67SE	317
Serum gln (μmol/l), 22% protein diet	spf/X	546+/-36	X/X	367+/-32	319

Note: Values expressed as mean +/- standard deviation.

this pH optimum shift and the decreased affinity for ornithine led to an activity of 10% of normal in the spf/Y liver. The heterozygote females showed a variation in activities between pH 6 and 10 that was intermediate between those of the affected males and the normal males, suggesting a dose affect due to the random, mosaic distribution of the abnormal gene in female liver(2). It was estimated that the spf mutation should be within 11 map units from the OTC gene on the X chromosome(148). Doolittle(312) reported a second mouse model of OTC deficiency also linked to the sparse fur defect called spf^{ash} (abnormal skin hair). The liver OTC activity in the spf^{ash} affected males varied from 5 to 10% of their normal mouse controls. This mutant will be discussed at the end of this section.

Spf/Y and normal X/Y mouse OTC were purified by using affinity columns of δ-PALO)(313), the immobilized transition state analog α-N-(phosphonacetyl)-L-ornithine. Both had subunit, monomeric molecular weights of 36,000. Using a rabbit antibody to the normal OTC, cross- reacting material in the spf livers was 120% of the controls in whole homogenate or in mitochondrial lysates. The pH maximum in spf OTC was between 9 and 10 and in the normal mouse liver was pH 8. The ratio of the activities of spf OTC at pH 9 over that at 8 was ~6 in liver homogenate or in mitochondrial lysates and the activity ratio of spf OTC at pH 9/normal OTC at pH 8 was 1.45. The specific activity of the normal mouse OTC at pH 8 was 300±30 μmol/min per mg pure OTC and was 400±30 at pH 9 for pure spf OTC. The Km for carbamyl phosphate in normal and mutant enzymes was 0.15 mM at all pHs, and the Ki for phosphate was 3.5 mM in both enzymes(314). The Km for ornithine as expected(89) varied with pH and the Km for the neutral form of ornithine was 0.06 mM at both pH 7 and 8 in the controls but was 0.07 mM at pH 7 and 0.39 mM at pH 8 in the spf enzyme(314). The Ki for norvaline, a competitive inhibitor vs. ornithine was only 0.1 mM in the normal enzyme compared to 0.8 mM in the spf enzyme confirming that there is a binding problem at the ornithine site(314). Characteristics of OTC from spf/Y and spf/X livers are compared with normals in Table 6-1(148,149,313–319).

Qureshi(315) showed that hemizygous spf males fed a standard mouse diet excreted 18.4 ± 5.1 μmoles of urinary orotic acid per mg urinary creatinine, compared with normal values of 0.72 ± 0.23, a 25-fold increase (Table 6-1). Heterozygous females excreted 7.38 ± 5.63 μmol/mg creatinine, with a range of 1.44 to 20.3, compared with normal female values of 0.54 ± 0.27. The wide range in heterozygotes is due to lyonization of the spf OTC gene in the female liver(2). The random plasma ammonia levels on a standard mouse chow diet (24% protein) were 2–4 times higher in spf/Y males than in normal or heterozygous females (Table 6-1)(317).

The molecular nature of this defect in the spf mouse was elucidated by Veres et al.(149). They made a mouse cDNA whose protein coding region was the same length (1062 bases) as in the rat and human. It was 96% homologous to the rat enzyme in sequence and 88% homologous to the human. The mRNA on Northern blots was 1650 base pairs, close in length to that in rat and human OTC. An antisense RNA probe was synthesized from the normal cDNA and bound to total RNA from wild-type and spf livers. RNAse cleavage was used to digest single stranded RNA and display any mismatches between normal and spf mRNA. A 1270-base-pair fragment of normal mRNA was protected by this cDNA. In the RNA of spf liver, a 1270 base fragment was protected but it was partially cleaved into 920 and 350 base-pair fragments. Further studies of the 5′-end of the molecule showed that the lesion was 350 bases from the AUG initiation site. A single C to A substitution was found at base 353, which changed the codon for histidine (CAC) to that for asparagine (AAC). The histidine is at position 85 on the mature monomer and at position 117 in pOTC. When spf cDNA was transfected into COS cells which contain no endogenous OTC, the OTC now present in the cells showed the same pH relationship as it did in the liver, proving that the defect did account for the abnormal pH response of the spf OTC. The histidine is very likely the ionizing residue with a pK of 6.6–6.8 identified previously(89). Why the pK of the Vm vs. pH curve should shift to 8.6 by substitution of an uncharged asparagine for a charged histidine is not apparent. Ohtake(318) found that the spf mRNA was of normal size (1.8kD), and amounted to 70% of that in controls. Nuclear OTC RNA was the same in amount as controls. Serum gln levels in heterozygous females overlapped with those of normal female mice(319) which means that affected females can not be identified by serum gln, or as noted above by plasma ammonia or urinary orotate.

Briand showed that the spf mouse has 150% of the normal amount of immunoprecipitable OTC protein(314). The level of translatable mRNA coding for pOTC is 58% of normal and the rate of enzyme synthesis in liver slices is 60% of controls(316). The molecular weight of pOTC in spf mice is 40 kDa, as in controls. Spf liver mitochondria process normal pOTC normally, and spf pOTC is taken up and processed by control mitochondria at the same rate as normal pOTC(316). A number of questions remain unanswered. Why does this single amino acid change in the spf sequence result in a lower level of mRNA and rate of enzyme synthesis? Why in the face of a lower mRNA and rate of synthesis should the final trimer be increased in amount? The speculation given by Briand was that the rate of degradation of the mature spf protein is slowed, but this has not been demonstrated, nor is there an *a priori* reason why this amino acid substitution should change structure enough to change degradation rates(316). The kinetic defects in liver OTC are also present in small intestinal mucosa(320,321) because the same gene is expressed in both tissues (Table 6-2).

The effects of the spf/Y trait on the mouse brain have been reviewed by Qureshi and Rao(322). We have compiled their reports and others published through 1999 in Table 6-3. These reports support the use of the spf mouse as a model of chronic hyperammonemic encephalop-

Table 6-2. OTC activities, OTC protein and pOTC mRNA in liver and small intestine of sparse-fur (spf) male mice

Tissue	Genotype	OTC activity[a] (% control)	OTC protein[b] (% control)	pOTC mRNA[c] (% control)
Liver	spf/Y	148	144	74
Small intestine	spf/Y	159	160	81
(% of liver)		15	15	44

Adapted from Dubois et al., 1988(321).
a. Assayed at pH 8 for normal X/Y controls and at pH 9 for spf/Y mice.
b. Western blots using polyclonal anti-mouse antibodies.
c. pOTC mRNA/poly(A)-enriched RNA by quantitative dot-blot analysis.

athy due to OTC deficiency. We have interpreted these findings listed in Table 6-3. The elevated plasma and brain ammonia levels lead to an elevated plasma gln and to increased brain gln via gln synthetase, which in brain is found primarily in astrocytes. The increased serum tryptophan (trp) levels are unexplained. A theory predicts that the increased brain gln is transported out of the brain by counter-transport with a neutral amino acid carrier, which leads to the increases in brain trp, tyr, phe and met. Increased brain ala and asp are assumed to result from increased ammonia combined with α-ketoglutarate to form glu, and then transamination from glu to ala and asp via pyruvate and oxaloacetate, respectively. Decreased brain arg and citrulline (cit) may reflect their low plasma levels due to impaired intestinal OTC levels and impaired conversion of gln to cit, which is then converted to arg in the kidney(327). Brain can convert cit to arg via argininosuccinate synthetase and lyase activities(111). Creatine synthesis in brain relies on arg from plasma, as does cerebral nitric oxide (NO) synthesis. NO in brain has not been reported in spf mice but NO synthase is decreased in these mutants(322).

The acquired, secondary deficiency of carnitine is worsened when the spf mice are treated with sodium benzoate, a means for increasing nitrogen excretion(328). The changes in brain energy metabolites are consistent with a block in glycolytic and citric acid pathways. The reduction in mitochondrial respiratory chain enzymes confirms this. The failure of transport of reducing equivalents into mitochondria is consistent with other reports of inhibition of the malate-aspartate shuttle, due to excess ammonia(322).

Peripheral-type benzodiazepine receptors are found on mitochon-

Table 6-3. Sparse-fur (spf) mouse as a model of chronic hyperammonemic encephalopathy due to OTC deficiency

Focus of study	Abnormalities detected
Serum amino acids	Plasma gln 133–551% of controls when plasma NH_3 is 244%(159–325) of controls. Serum trp increased 135% of controls; all other amino acids decreased, including arg, orn, cit.[a]
Brain amino acids	Gln 197% of controls when NH_3 is 184%. Tyr is 217%, phe 223%, met 166%, ala 142–180%, trp 128–200%, his 240%, GABA 130%, asp 160%, arg 35–60%, orn 59–150%, cit 67–78% of controls. Creatine decreased also.[a]
Liver, muscle carnitines	Total, free, short and medium chain acyl-carnitines are 40–70% of controls.[a]
Brain energy metabolites	↑ glucose, lactate, α-ketoglutarate; ↓ pyruvate, glutamate and ATP. NADH/NAD ratio ↑ in cytosol, ↓ in mitochondria. ↓ CoA, acetyl CoA.[a]
Respiratory chain enzymes	Cytochrome c oxidase activity ↓ to 52–64% of controls; ↓ NADH-cytochrome c reductase, succinate cyt. c reductase, cyt. c oxidase in synaptosomal mitochondria.[a,c]
Peripheral type benzodiazepine receptor densities	↑ number of binding sites in astrocytes of brain and in liver, kidney, testis.[a]
Brain serotonin dihydroxyphenylacetate and its receptors	↑ trp, 5HIAA and serotinin; ↑ norepinephrine, dopamine, ↓ monamine oxidase- A(MAO) activity but ↑ MAO-B activity in cerebellum and brain stem. Number of binding sites for $serotonin_2$ receptors ↓ but ↑ for $serotonin_{1a}$ receptors.[a]
N-methyl-D-aspartate-type glutamate receptors	↑ quinolinic acid, an NMDA receptor binder; loss of medium-spiny neurons due ? to excitotoxic over-stimulation of NMDA-type glutamate receptors.[a,b]
Brain cholinergic system	↓choline acetyltransferase(ChAT) activity, ↓ high affinity choline uptake, ↓ β-nerve growth factor, ↓M_2 cholinergic receptor densities (pre-synaptic) but ↑ M_1 receptors (post-synaptic) and ↓ ChAT-positive neuronal cell counts in cortex and septal area.[a]
Glutamate and its receptors	Glu ↓ only in cerebral cortex. NMDA binding sites ↓ in pyramidal neurons of fronto-parietal cortex and ↓ in neuronal dendritic spine density in this area.[a,d,e]
Na^+-K^+-ATPase in brain	↑activity in cerebral areas, plus ↓ ATP levels.[a]

Results limited to adult spf/y males.
a. reference 322. b. 323. c. 324. d. 325. e. 326.

drial membranes and in the brain on astrocytic mitochondria. The reasons for increased densities of these receptors or the consequences are not known. The increased brain levels of trp and its metabolite, 5-hydroxyindoleacetic acid (5HIAA) have been seen in other models of hyperammonemia. Serotonin, a neurotransmitter, is increased by this rise in precursor levels(322) and remains high because monamine oxidase-A (MAO-A), the enzyme which catabolizes serotonin, is decreased. The high serotonin levels down-regulate serotonin$_2$ receptor binding site numbers but strangely up-regulate the number of serotonin$_{1A}$ receptors. Quinolinic acid, another trp metabolite, is also increased in spf mice and in CSF of children with genetic urea cycle deficiencies(323). This excitotoxin binds to the N-methyl-D-aspartate (NMDA) sub-type of glutamate receptors. Overstimulation of these receptors has led to calcium-mediated neuronal death(325). The decrease in NMDA binding sites in spf mouse cortex may be due to injury to the neurons in this area. Moreover, decreased medium-spiny neurons were found in the striatum of spf mice(326). From 35 days of age, spf/Y mice showed an elevated calcium dependent release of endogenous glu from synaptosomes and a reduced uptake of glu into synaptosomes, along with reduced MK-801 binding to the NMDA sub-type of glu receptors. Treatment with acetyl-L-carnitine reversed these defects toward normal, possibly by repletion of the ATP needed for transport of glu into synaptosomes(329).

A number of abnormalities occur in the brain cholinergic system (Table 6-3). These defects may help explain the behavioral abnormalities in spf mice(322,329) and the behavioral and cognitive defects in children with urea cycle disorders(330). Decreases in glutamate levels and in NMDA binding sites for glutamate are postulated to result from loss of these cortical and hippocampal neurons or down-regulation of the receptors due to excessive glutamate or quinolinic acid stimulation of the receptors(322,331). In chronic hyperammonemic animal models of other kinds and in humans with hepatic encephalopathy and hyperammonemia a concensus is arising that excessive glutamate in the neural cleft overstimulates the NMDA receptors, causes down-regulation of glu receptors and loss of neurons(331,332).

Lastly, excessive ATPase activity may play a role in depleting ATP levels in the various cerebral regions of spf mice (Table 6-3). How this activation occurs is mostly speculative(322).

Treatment of spf/Y males with ornithine reduced their high orotate urinary excretion, indicating that residual OTC activity at pH 7.4 in mitochondria can be maximized[323]. This ornithine effect was confirmed by treating spf/Y mice with 5-fluoromethylornithine, an inhibitor of ornithine aminotransferase. The subsequent increase in endogenous liver and brain ornithine levels reduced blood and tissue ammonia concentrations and urinary orotic acid excretion to near normal[334]. In humans with OTC deficiency supplements of citrulline or arginine also reduce urinary orotic acid and plasma ammonia, presumably by serving as a source of liver ornithine. Qureshi[319] found that the lower limits of excretion of orotate in the spf heterozygote females overlapped with the upper normal values of the wild type females, a result of lyonization whereby many spf/X female livers contain almost normal amounts of wild-type OTC[2]. Spf/Y males respond as do humans with OTC deficiency to feeding sodium benzoate by reductions in plasma ammonia[335] and urinary orotate[319]. The hepatic and renal synthesis of hippuric acid from benzoyl-CoA and glycine leads to waste nitrogen elimination as urinary hippurate by channelling ammonia through glutamine and eventually to glycine. Secondary carnitine deficiency occurs in spf/Y males[322] and this is aggravated by benzoate; feeding carnitine restores depleted CoA stores[324].

Sparse Fur^{ash} Mutant Mouse

The characteristics of the sparse fur-ash (spf^{ash}) mouse model of OTC deficiency, discovered by Doolittle et al.[312], are listed in Table 6-4. The mechanism by which the spf^{ash} defect occurs has been elegantly elucidated by Rosenberg[336] and Hodges[337]. They found that the spf^{ash} male has an activity 7.8 percent of controls, the cross reacting material is 5.8 percent, the pI of 7.55 is the same as controls, the pH optimum of the enzyme is normal (7.8 to 8) and the K_ms for ornithine (l-l.3 mM) and carbamyl phosphate (l.2–l.3 mM) are the same as controls. These results confirm those of others[338]. An Ouchterlony diffusion plate showed lines of identity between the spf and normal OTC and they were similar on immunoelectrophoresis. However, when polysomal RNAs were isolated and translated in the rabbit reticulocyte lysate system, the spf^{ash} male showed two bands of 36 and 37 kDa at a ratio of 3:2. When there was an equal amount of RNA added to the translational assays, the spf^{ash} extract contained only 15 to 20 per-

Table 6-4. Characteristics of the sparse-fur-ash mouse model of OTC Deficiency

Characteristics		spf		wild-type	references
pOTC cDNA change		CA-T(change of donor splice site, nt 386 on exon 4)		CG-T(nt 385-7)	337
pOTC amino acid change		H 129(active)		R 129	337
pOTC cDNA length		~1700, ~1700 + 48nt		~1700	337
pOTC mRNA length		1650, 1698(cryptic splice donor site in intron 4)		1650	337
OTC trimer size		108kD		108kD	336
OTC monomer size		36kD		36kD	336
OTC activity(μmol/hr/mg protein, at pH 7.7–8, CP 5mM, orn 5mM)	ash/Y	2.65	X/Y	54+/-1.1	338
		2.29+/-0.06SE		52.1+/-2.6SE	321
		5.7+/-1.6		73.2+/-8.5	336
	ash/X	40.7+/-22.1		—	336
V_{max} liver(μmol/hr/gm wet wt.), (pH 7.7, CP 10mM, orn 0.02–10mM, fed 24% protein mouse chow)	ash/Y	1060+/-113	X/Y	16200+/-1040	340
K_m liver (mM)					
CP(pH 7.7, orn10mM, CP 0.08–10mM)	ash/Y	0.55+/-0.36	X/Y	0.45+/-0.22	340
Orn(pH 7.7, CP10mM, orn 0.02–10mM)	ash/Y	0.36+/-0.08	X/Y	0.42+/-0.19	340
pH optimum		7.8–8		7.8–8	336
		7.95		7.95	340
pI (isoelectric point)		7.55		7.55	336
Cross-reacting material (% control)		5–10, 10			321, 338
		5.8+/-0.8			336

Characteristics		spf		wild-type	references
Translatable mRNA (% control)		10 15–20			316 336
pOTC mRNA (% control)		10, 13			321, 337
Orotic acid excretion, urinary(μmol/ mmol creatinine, fed 24% protein diet)	ash/Y ash/X	321+/-536 124+/-241	X/Y X/X	32.4+/-11.6 7.7+/-2.3	340 340

Values expressed as mean+/- standard deviation, except when standard error, SE noted.

cent of the normal translatable mRNA. In normal mitochondria, mutant pOTC was taken up and processed via an intermediate OTC of 37 kDa and a normal monomer of 36 kDa resulted. When spf^{ash} male translation products were examined carefully, however, a larger intermediate OTC and a larger monomer were also present(336). In the mitochondria isolated from spf^{ash} mice the OTC was the same as controls, 36 kDa.

The initial hypothesis was that there was an error in some of the intron-exon splicing which led to an increased size of the mRNA for the spf^{ash} gene and that the resulting enlarged pOTC could be taken up by the mitochondria and cleaved but could not be assembled into trimers and was subsequently degraded. Hodges and Rosenberg confirmed this hypothesis(337). They made a cDNA construct from normal mice containing 40 base pairs of 5′-sequence plus the pOTC cDNA and 263 base pairs of 3′-sequence. Using this probe they found the total spf^{ash} mRNA was only 10 percent of controls. When an antisense cDNA was hybridized with total spf^{ash} RNA and then digested with Sl nuclease they were able to find a normal mRNA, and also a second mRNA with a defect at the junction between exon 4 and intron 5. Analysis of PCR products in this region found two defects. The first was a G to A change in the final base of exon 4 from CGT to CAT, an arginine to histidine mutation. Loss of the GT recognition pair for a donor splice site resulted in a pre-mRNA that was 48 base pairs longer and coded for a protein containing 16 extra amino acids. The splicing mechanism made 5% of the normal-sized histidine-coding mRNA and 95% of a nuclear pre-mRNA spliced at a cryptic splice site 48 bases into the intron. Most of the mis-spliced mRNA was degraded in the nucleus but some elongated mRNA escaped into the cytoplasm, and programmed synthesis of an elongated pOTC. The longer pOTC was taken up into mitochondria, cleaved and yielded a longer monomer that could not be assembled into trimers. The histidine-containing monomer could be assembled into an active trimer. Even wild type mice have a problem splicing in this area of the gene and a small percent of errors occur so as to join exon 3 to exon 5[(337)]. Overall, therefore, the spf^{ash} defect consists of a mutation which creates two mRNAs in low amounts, a his-129 mutant which can be assembled into functional enzyme at 5 percent of normal levels and a 16-amino acid elongated monomer which cannot be assembled into active enzymes and is degraded. Ohtake confirmed that the spf^{ash} nuclear RNA was much reduced and that translatable mRNA was only 5–10% of normal[(318)]. Briand also confirmed that there was 10% translatable mRNA in spf^{ash} livers and recognized the normal and larger pOTCs with but one normal 36 kDa monomer[(316)].

In the spf^{ash} heterozygotes Qureshi[(319)] (Table 6-4) found there was overlap between the wild type and the heterozygote female OTC activities so there was no ability to discriminate between them. He confirmed that the affected males with the spf^{ash} defect had 4.1 percent of normal liver activity and the female homozygotes had 3.5 percent. The female heterozygotes averaged 37.8 units per mg liver protein compared to 96.6 in the wild type female. The serum glutamine levels correlated significantly with the liver level of activity in both spf and spf^{ash} mice[(319)]. Serum ammonia levels did not discriminate between normal and heterozygote females in either group. Therefore, it appeared that the only way to tell an spf^{ash} heterozygote from a normal heterozygote was to breed the female with a normal male and look for the phenotype of sparse fur in some male offspring.

Cohen[(339)] studied citrulline synthesis in the mitochondria of spf^{ash} mice and found that the mitochondria contained 33% more mitochondrial protein per g of liver than preparations from normal liver. Yet total liver protein/g liver was only 12% higher in spf^{ash} mice. OTC was 6% of normal in units/g liver while CPS activity was normal. The ratio of OTC/CPS in normal liver is 39 and in spf^{ash} livers only 2.1. Other mitochondrial enzymes were increased in activity, glutamate dehydrogenase by 16%, citrate synthetase by 28% and β-hydroxybutyrate dehydrogenase by 66% in spf^{ash} livers. Surprisingly, spf^{ash} mitochondria respiring on succinate with 10mM ornithine produced citrulline at a rate of 25 nmol/min compared to rates of 33–36 in normal mitochondria. Maximal carbamyl phosphate synthesis in uncoupled normal mitochondria incubated with acetylglutamate was 73 nmol/min per mg protein with 64 nmol/min per mg being converted to citrulline. Spf^{ash} mitochondria treated the same way gave carbamyl phosphate values of 46 produced and 44 utilized. Both results do not correlate with an OTC activity 5% of normal, a ratio of OTC/CPS of 2, and with the orotic aciduria which occurs in these hemizygote males *in vivo* due to underutilization of carbamyl phosphate in citrulline synthesis. Overall, this study suggested to Cohen et al.[(339)] an adaptation to OTC deficiency in the spf^{ash} mouse by increasing the mitochondrial number per cell.

We have found that normal and spf^{ash}/Y males and homozygous spf females showed induction of their low OTC activity levels by feeding a high protein diet (Table 6-5)[(340)]. This phenomenon is well known in normal rats[(341,342)]. In rats, protein and certain amino acids probably induce OTC directly and also induce by releasing glucagon, which together with corticosteroids, synergistically induces OTC activity[(343)]. OTC mRNA is also induced in rats by feeding a high protein diet[(345)]. Induction of mouse liver OTC has not been studied in depth but is likely to be under similar controls as in the rat[(20,21)]. The induction of spf liver OTC by high protein feedings (Table 6-5) means that the responses to inducers of the OTC enhancer and promoter regions are intact even though the structural gene is mutated.

Gene Therapy of Mutant Mice

Once these two mouse models were characterized as to gene and enzyme defects and to pathologic consequences, the knowledge was used by many investigators for development of gene therapies and pharmacologic treatments of OTC deficiency. One of the earliest and most successful approaches was to create a *transgenic* spf or spf^{ash} mouse harboring the rat or human OTC cDNA. Cavard et al.[(345)] made a construct of the SV40 early promoter on the 5′ end of a rat OTC cDNA and an SV40 3′ polyA tail and injected it into the male pronucleus of homozygous spf^{ash} mouse oocytes fertilized by normal mice. They implanted the 85 surviving treated eggs in the uteri of

Table 6-5. OTC activities in sparse-fur-ash vs. normal mice livers after feeding mice 15% or 60% casein diets for seven days[a]

Genotype	X/Y		Ash/Y		Ash/Ash		Ash/X
Casein diet	15%	60%	15%	60%	15%	60%	60%
	(units per 100g of mouse)[b]						
Mean	76600	129000[c]	5600	11800[c]	4450	7080[c]	58400
Std. Dev.	13900	21600	2490	3440	1680	1980	26500
Variance	18%	17%	44%	29%	38%	28%	45%
n	5	8	13	9	13	16	7

a. Snodgrass, unpublished data(340).
b. μmol/hr/g liver multiplied by g liver/100g of mouse.
c. p<0.001 vs. same genotype on the 15% casein diet.

pseudopregnant mice and obtained 21 pups. One litter of two males had one male whose fur was normal and had the rat OTC cDNA incorporated into its genome. Offspring had 50% of normal liver mRNA levels and 80% of normal OTC activity in liver and small bowel mucosa(345).

Using a construct of 1.4 kb of the 5′ flanking region of the rat OTC gene, a 1.2 kb cDNA of the rat OTC protein coding region and 2.5 kb of the 3′ flanking region of the rat growth hormone gene, Murakami et al.(346) injected this 5.1 kb construct into fertilized eggs of C57BL/6 mice and obtained 17 out of 125 offspring which carried copies of rat OTC. The transgene in liver was only a few percent of the endogenous OTC mRNA but 10–100% of the endogenous OTC in small intestine. Intestinal OTC activities were increased from 114 to 148% of controls(346).

Jones et al.(347) made a construct containing 750 bp of mouse OTC 5′ promoter sequences plus the first 17 bp of translated sequences coupled to an 1160 bp human OTC cDNA and an SV40 splice donor/acceptor sequence and polyadenylation sequence 3′ to the human OTC. Transgenic mice were created with this vector by injecting it into the male pronucleus of homozygous spf female oocytes fertilized by hemizygous spf males. Two pups out of 19 survivors had normal fur and weight and bore the human OTC transgene, inserted at a single site in a head to tail fashion in founder pup FO5. In the other founder pup FO4 three sites of transgene insertion were found, containing one, a few and a high number of copies. Both progenies showed high levels of transgene in small bowel and very low levels in liver[347]. Hepatic OTC activities were the same as spf mice but a 17-fold increase in intestinal OTC activity was found. This combination of 9300 pmol/hr/mg protein in liver and 2380 in small bowel compared to spf mouse levels of 11900 and 136 respectively restored urinary orotic acid levels to near normal. Thus it was apparent that the 750 bp mouse promoter region did not contain elements coding for liver-specific expression of human OTC[347]. Moreover the normalization of urinary orotic acid excretion means that the small bowel OTC utilized carbamyl phosphate to make citrulline and thereby decreased production of gut orotic acid[349]. The OTC activity in spf liver is 15% of normal and this appears to be enough to keep mitochondrial CP from being shunted to orotic acid synthesis as long as the gut and kidney supply the liver with arginine[327].

Shimada et al.[348] created transgenic spf^{ash} mice using the construct of Murakami[346]. The liver OTC of hemizygous transgenic males was twice that of spf males and the small intestine activity was six times higher, the values being 12% of normal mice in liver and 27% in small bowel. These transgenic spf males had normal plasma citrulline levels and urinary orotic acid levels of 907 ± 141 nmol/mg creatinine compared to normal mice levels of 477 ± 303. They also had normal carbamyl phosphate levels in liver and small intestine[349]. The intestinal OTC activity probably corrected the low liver ornithine levels in spf^{ash} males (245 ± 81 nmol/g vs. 338 ± 155 in control mice) by correcting fasting liver arginine deficiency[327,349]. Thus the transgene experiments prove the feasibility of correcting OTC deficiency in the spf and spf^{ash} models by creating transgenic individuals, but the efficiency of DNA insertion in the genome of injected fertilized ova is still quite low.

Because of ethical problems in ever applying this germ-line gene therapy technique in humans, almost all investigators turned to *somatic* gene therapy of liver and intestinal cells[350,351]. The first successful attempt was carried out by Stratford-Perricaudet et al.[352] who created a recombinant adenovirus which contained the rat OTC cDNA under the control of the viral major late promoter (MLP). The plasmid construct (PMLP-OTC) consisted of 455 bp of the 5′ end of the adenovirus-5 genome, deleted for E1 genes necessary for late gene expression, the adenoviral MLP, a complete cDNA of the rat

OTC coding region and a portion of SV40 genome on the 3′ end for RNA maturation signals. The PMLP-OTC was ligated to adenovirus 5 d1327 to form AdMLP-OTC, the infective unit.

This AdMLP-OTC recombinant virus infected a packaging cell line which supplies the deleted E-genes (293-cells) so that high titers of virus can be grown. This infection resulted in OTC activities of 2.55 μmol citrulline/hr·mg cell protein in these cells compared to no OTC activity in mock-infected cells. This virus construct was injected intravenously into spf^{ash} mice at one day after birth with 20–40μl at a dose of 10^9 pfu/ml. At one month liver OTC activities ranged from 6.4–61.5 compared to 70 ± 5.7 units/mg protein in control mouse liver and 3.6 ± 1.3 in spf^{ash} liver. By two months the levels were 2.4–5.8 and at 15 months were 0 and 38.4 in the two surviving mice. Urinary orotic acid excretion in the two mice tested at 15 months was 7.1 and 1.5 μmol/mg creatinine compared with spf^{ash} values of 20.2 ± 5.5 and with normal controls of 0.5 ± 0.2. Viral OTC by PCR at 15 months was present in the livers of the two mice. The probable reason for the persistence of the viral OTC DNA is the development of immunologic tolerance by the one day-old mice[(352)].

Morsy et al.[(353)] were able to infect cultured spf hepatocytes with a construct AdSRαhOTC consisting of the first adenovirus 5 sequences with SRαhOTC inserted in the E1 region between 452 and 3328 bp. The SRα promoter is a combination of the SV40 early region promoter linked to human T cell leukemia virus-1 enhancer sequences. Reporter gene vectors in replication-defective adenovirus OTC constructs infected, at a dose of 100–299 viral particles per cell, 100% of cultured liver cells *in vitro*. One week after infection with multiplicities of infection (moi) from 1 to 200, OTC activities rose from 26 to 151% of those in wild-type mouse hepatocytes. Primary human hepatocytes were isolated and cultured from a severely OTC deficient patient carrying an exon3, Taq1 mutation. Four days after infection with AdSRαhOTC at a dose of 200 moi the cells showed a non-quantified OTC activity. Newborn spf^{ash} mice injected with AdSRαhOTC (2×10^{10} pfu) corrected orotic acid levels in 24 hours and this correction lasted 15 weeks[(353)].

The choice of adenoviruses was based on their ability to infect hepatocytes *in vivo* in their quiescent state and to be rendered non-replicative by deleting the E1 gene which activates late gene expression and thus blocks assembly of adenoviral particles[(350)]. However these defective adenoviruses can express early gene products and can be *lytic* under certain conditions or can elicit a cellular immune reaction. Moreover the early gene products elicit blocking antibodies that make retreatment ineffective. Combining adenovirus with receptor-mediated gene delivery increases uptake and expression of DNA by slowing degradation of the DNA in endosomes(354). An adenoviral vector remains independent in the nucleus and does not integrate into the host genome. It is gradually lost with cell division or in a liver cell that does not divide is lost by degradation by nucleases. An adenovirus is a large (~38 kb) double-stranded DNA. The genes expressed early in its life cycle are required for DNA replication and viral propagation. Deletion of early gene E1-A means the virus cannot regulate the genes needed for replication. The vector is grown in a helper cell line (293-line cells which supply the missing E1-A gene) to a titer of 10^{10-11} viable particles/ml. The maximum size of a therapy gene insert is 7–8 kb, adequate for OTC and its promoter(355). Hepatic cells cannot supply E1-A gene function as do helper cells. Adenoviral vectors deliver recombinant genes to lung, muscle, blood vessels and brain as well as liver. Cellular immunity to viral antigens in E1-deleted adenoviruses not only prevented repeated treatments but also led to destruction of genetically modified hepatocytes(356). Adult mice mount a cellular immune response to infected liver cells but newborn mice exposed to an adenovirus vector before immunologic maturity develop tolerance to the viral antigens. Humans only develop tolerance during 14–18 weeks of gestation(357).

Morsy(358) used the AdSRαhOTC vector and injected 10 μl *into the livers* of newborn spf^{ash} pups at doses of $1–5 \times 10^8$ pfu. Almost 100% success was achieved in lowering urinary orotic acid excretion and the reduction was 70% at 24h and 71% at 3 weeks. Phenotypic correction of hair growth occurred in 42% by 2 weeks, usually associated with a 24% increase in weight. OTC activity at 3 weeks was 54% of that in wild-type control liver and 44% of that in control intestine. Intravenous injection of the vector did not increase OTC activity in liver or intestine. The vector strongly increased the hOTC protein in spf or COS-7 cells, but the OTC activity in COS-7 cells was much higher than that in hepatocytes. Equimolar mixtures of dissociated and reassociated mutant and wild-type proteins resulted in 33% reductions in activities compared to the immediate assay of the mixture. This was interpreted as formation of less active *heterotrimers* of OTC (two wild-type human OTC + one spf OTC monomers) as Wente and Schachman demonstrated for aspartate transcarbamylase[359], known as a *dominant negative effect*. It means that the corrective power of transfecting with normal hOTC would be greater in patients with low levels of mutant OTC monomers, i.e. more severe enzyme deficiency.

Kiwaki et al.[360] created a vector AdexCAGhOTC using an adenoviral-5 deletion of E1A, E1B and E3, followed by the CAG promoter (a CMV-1E enhancer plus a modified chicken β-actin promoter), an hOTC cDNA inserted into the E1 site, and a rabbit β-globin 3′ flanking sequence with a polyA signal. Intravenous injection of this vector into spf^{ash} mice at 10^9pfu/mouse gave an average liver OTC activity after 7 days of 87 ± 32% of wild-type levels. Urinary orotic acid levels at 7 days were within normal mouse levels. Neutralizing antibodies at 1:20–1:200 developed and liver sections of a mouse who died after a dose of 3.5×10^9 pfu/mouse showed focal necrosis and inflammation. Sixteen other mice showed minimal sites of necrosis. Doses of 0.5×10^9 pfu/mouse yielded 41 ± 7% of control OTC activity. Primary human hepatocytes from a liver transplant female patient with an R23stop codon mutation and 6% overall OTC activity were infected with this vector and at 10 moi achieved normal human OTC levels.

Ye et al.[357] tried and failed with a number of E1 deleted adenovirus constructs in spf mice, using hOTC cDNA driven by the CMV-enhanced β-actin promoter. The most successful vector used a 5′ adenoviral-5 sequence of only 0–1 map units, then a strong constitutive viral enhancer/promoter from CMV, followed by the mouse OTC cDNA in place of viral E1 gene, an SV40 polyA site, then the viral Ad-5 sequence 9–100 map units with the E2A gene made temperature sensitive by a single base pair substitution (ts 225) and a sub360 E3 segment, all together named H5.110CMVmOTC. Intravenous infusion of this vector at 2×10^{11}pfu more than doubled OTC activity over C3H controls. Histochemical analysis showed all cells transfected. Activity was seen in most cells on day 28. Urinary orotate decreased to 10% of spf/Y controls and returned to control levels by 70 days. Spf^{ash} males were reconstituted to the same degree. Inflammation was reduced and late viral genes were hardly expressed. Serum glutamine was normalized by 20–40 days in treated spf/Y mice but citrulline remained low, probably due to failure of small intestine reconstitution of OTC. The mechanism for elimination of this adenoviral vector is not yet understood.

When this vector was infused into spf/Y mice which were then challenged with NH_4Cl (10mmol/kg) over 1–28 days, partial protection from seizures and biochemical abnormalities was achieved by 24h and complete protection by 48h[361]. Protection declined by 14 days and was gone by 28 days. OTC levels in liver rose to near control levels through 14 days but were the usual spf/Y level of 5% by 28 days. Batshaw et al.[362] also infused intravenously H5.110CMVmOTC into spf/Y mice at a dose of 4×10^{12}pfu/kg and carried out measurements of rates of ureagenesis using $^{15}NH_4Cl$. At 7 and 14 days after infection there was a 2–2.5-fold increase in ^{15}N incorporation into urea, raising the rate from the spf level of 60% of control mice to 100%. Transfected spf/Y liver OTC activities at 7 days reached 77.5 ± 8.4 μmol/h·mg protein vs. 85.2 ± 12.2 in C3H controls. At 14 days activity was ~15 units. Orotate excretion fell to normal levels at 7 and 14 days. Infusion of an anti-CD4 monoclonal antibody along with repeated injections of the Ad.CMVmOTC vector diminished the immune response and prolonged persistence of OTC activity in spf liver.[364]

When this vector was infused into baboons via the hepatic artery, at doses of 6×10^{11} pfu/kg, mild inflammation was found in the liver at day 61, with vector DNA found in liver at days 29 and 61, but none in gonads[364]. A phase-1 trial in humans was approved by the Food and Drug Administration using H5.110CMVhOTC. Details of human therapy with this vector are found in Chapter 8.

Hope for success using adenoviral vectors to deliver stable corrective OTC genes to spf mice has been raised by the report of DelloRusso et al. in dystrophin deficient *(mdx)* mice[723]. A "gutted" adenoviral vector was made which deleted all viral sequences except those needed in cis for viral replication and packaging. This vector containing a full length dystrophin gene efficiently transduced muscles of 1-yr-old *mdx* mice and showed functional dystrophin a month later. Another advance was the inclusion of a DNA transposon in a gutted adenoviral vector that caused stable genomic integration of a human coagulation factor IX into mouse liver which had been induced to regenerate[724]. Immune responses to these new vectors are a concern.

Retroviral vectors need cell replication, rarely can be grown in titers high enough for efficient *in vivo* applications and carry potential risks of insertional mutagenesis or development of replication competent virus[350,351]. However, Grompe et al.[365] were the first to use a recombinant retrovirus to transduce OTC. The vector consisted of the long terminal repeat (LTR) of retrovirus ΔN2, then 40 bp of 5′ untranslated hOTC, the protein coding region of hOTC, and 100 bp of 3′ untranslated hOTC followed by LTR. One clone ΔN2OTC-8 could be grown to 10^6 particles/ml in NIH-3T3 cells and was used to infect cultured spf and spf^{ash} hepatocytes. In spf^{ash} cells the integrated copy number was 0.5–2 copies per genome. The mRNA levels of hOTC in spf cells averaged 116% of OTC in adult human liver (33–333%) and were stable for 10 days. OTC activity in spf hepatocytes at day 15 was increased to 344 μmol/hr·mg protein compared to 8 units in controls but still was only 3% of wild type levels. Spf^{ash} OTC activities doubled by day 8 over controls. It was hoped that retrovirally induced hepatocytes might be transplanted into liver *in vivo* in the future.

However Podevin et al.[366] found that retroviral-mediated transfer of a marker gene in spf^{ash} mice never exceeded 20 liver cells per cm^2 in spite of two-thirds hepatectomy to induce liver cell division and perfusion of the liver with retroviral vector after 24–48 hours. Mortality was 8/20 mice in spite of low protein diets and glucose feedings. The use of retroviral vectors *in vivo* has been abandoned by most investigators until these problems are solved.

One other animal model of OTC deficiency is the juvenile visceral steatosis (JVS) mouse that develops a fatty liver after birth and at 25 days shows a decrease of all five urea cycle enzymes to 10–40% of control levels[367]. This loss of activity is due to a reduced level of OTC transcription which leads to low mRNA levels[367] and to genetic deficiency of high-affinity L-carnitine receptors and uptake into JVS hepatocytes[368]. Albumin and serine dehydratase mRNA are also severely reduced. The defects in transcription are corrected by treating the mice with high dose carnitine[367,368] that is taken up by low-affinity carnitine receptors. Carnitine deficiency explains the hyperammonemia, urea cycle deficiencies and impaired weight gain, but not the fatty liver.

Commentary. Research workers on OTC deficiency are fortunate to have two spontaneous mouse models, the sparse-fur (spf) and spf^{ash}

mice. Their gene defects are X-linked, the gene defects are known, the detailed kinetics and characteristics of the enzymes are known and the severity of the deficiency is moderately severe, with 5–20% of normal mouse liver OTC activity which allows survival of affected males and their use in breeding. They have served as models for chronic hyperammonemia, leading to improved understanding of the deleterious effects of ammonia on the brain and to their use as test animals for therapy of hyperammonemia. A number of reports describe successful germ-line therapy with normal OTC cDNA constructs, creating transgenic mice. Somatic cell therapy of liver and gut has been achieved using modified adenovirus vectors that led to human trials.

7

Clinical and Laboratory Findings in OTC Deficiency

We have collected most of the reported cases of OTC deficiency between 1962, when the first case was reported, through 2002. We searched Index Medicus(Pub Med)and Reference Update(ISI) for the keywords ornithine transcarbam(yl)(oyl)ase, ornithine carbam(yl)(oyl) transferase, and other keywords for elements of the urea or ornithine cycle in articles written in English, German, Spanish and French. Articles written in Japanese were not included. Copies of each article were reviewed and tables were constructed listing in order of year of publication all female cases (Table 7-1), neonatal male cases (onset 0–28 days after birth)(Table 7-2) and late-onset male cases (Table 7-3)(>28 days). We included clinical findings reported at onset of the illness and up to the time of diagnosis. Symptoms we included were (a) vomiting; (b) decreased level of consciousness from lethargy to coma; (c) focal or generalized seizure activity; (d) abnormal behavior which persisted or recurred; (e) avoidance of protein-containing foods or intolerance of infant formulas; (f) delayed growth (<3% of predicted height or weight); (g) delayed development (IQ or other cognitive tests below normal); (h) incoordination (usually ataxia); and (j) some form of persistent central nervous system (CNS) deficit. We also included in the list of clinical findings (i) a positive family history of OTC-type illness, defined as proof of OTC deficiency in the mother, siblings or first-order relatives, or a history of at least two neonatal male deaths in siblings without a known cause. Cases in the literature were not included in tables 7-1, 7-2 or 7-3 if no clinical symptoms were given. One late onset case was included with only a positive family history.

The age at onset of symptoms compatible with OTC deficiency and the age at death were taken from the case history. When no definite age was given for onset or death a "?" is appended in the tables. If the death was noted as 7 days after onset it is recorded as D/+7d. The blood or plasma ammonia level first measured or at onset is expressed as the fold-increase over the author's stated upper limit of normal. We did not use *peak* ammonia values because they often occurred after the diagnosis was established or days after onset of symptoms. If no normal values were given we used 50 μmol/l or 85 μg/dl as the generally accepted upper limit of normal by most methods. Similarly we expressed urinary levels of orotic acid (± orotidine) as fold-increases over the stated upper normal by the author's method. For those upper limits not stated we used an average value (n=9) from the literature of 3.5 μmol/mmol creatinine. If data were reported, acid-base status was defined as *respiratory alkalosis* if the arterial pH was >7.45 and the pCO_2 was <35 mmHg; as *metabolic acidosis* if the pH was <7.38,the pCO_2 <35 and HCO_3 <21mmol/l; and as *metabolic alkalosis* if the pH was >7.45,the pCO_2 >45 and HCO_3 >30 mmol/l.

Liver OTC activity was expressed as % of the normal mean value, assayed at the human pH optimum range of 7.7–8.3. When the pH limits were not given, we used the author's calculation of % normal. K_m values derived from patients' liver assays were compared with the author's normal mean values and expressed as mmol/l (mM). Cross-reacting material (CRM) or Western blots, which reflect liver OTC protein concentration, were measured by an immunoassay and we expressed them as percent of the author's control liver values. The gene defects are described in greater detail in Table 5-1. In Tables 7-1, 7-2 and 7-3 we give the changes in pOTC amino acid sequence (i.e., R277W) or change in intron sequences (i.e., g→a, nt 540+1, intron 2).

OTC Deficiency in Females

The first patients with documented OTC deficiency, two female first cousins, were reported by Russell in 1962[369], although only one was proven at the time of publication by assay of OTC in liver to have the

Table 7-1. Clinical and laboratory findings in reported female probands with OTC deficiency, 1962–2002

Case No.	Clinical findings[1]	Age at onset	Alive (A) or age at Death (D/ y)	Blood or plasma NH3[2]	Acid-base status[3]	Orotic aciduria[2]	Liver OTC[4] activity % of control[4]	Km (orn)[5] vs. control mM	Km (CP)[5] vs. control mM	CRM[6]	Gene defect	Reference
1.	a, b, d, f, g, i	5w	D/6y	6.3		40	4.8					369, 372
2.	a, b, d, f, g, h, i	3y	D/6y	3.4			7.1					369, 372
3.	b, e, i	9y	A	0.89	N	5						370, 373
4.	a, b, c, e, f, i	3w	A	19	N		8					371, 373, 374
5.	e, i	10m	A	2.6			40	(mother of case 4)				371, 373
6.	a, b, d, e, f, g, h	7m	D/18m	12	RAl		4					375
7.	a, b, f, g, h	14d	A	2.5	N	127	33					377
8.	a, b, c, d, g, i	5m	A	3.8	N		11					379
9.	a, b, e	8m	D/10m	3.6	N	33	69	1.25/1.45	6.65/1.67			380
10.	a, b, d, f, g, h, j	3.5y	A	7.5	N	12.5	5.2					381
11.	a, b, g, i	1m	A	2.3			12	0.25/0.46	2.1/0.54			382
12.	a, b, c, d, i	1m	A	2.3	N		36					383
13.	a, b, d, f, h, i	14m	A	7.5	RA1		23					384
14.	a, b, c, d, g, h	1m	A	2.5	MAl		19					384
15.	a, d, e	7d	A	1.6	N		26					384
16.	a, b, e, f, g, i	1m	A	1.4			1.4					385
17.	a, b, h	5y	D/+3d	>11.8			3					386
18.	?(ketosis)	4y	A	4.1			11					387
19.	b, c, e, g, h	6d	A	8.3		462	13.4	0.41/0.56	0.18/0.45			388, 389
20.	b, e, i	6d	D/3.9y	4.2			5	0.69/0.37	0.26/0.27			390
21.	a, b, c, i	5d	A	1.4	MAc	381						391
22.	a, b, e, g, i	2.9y	A'	1.7		23	13.4					392
23.	b, d, e	12y	D/+3d	8		240						392

Case No.	Clinical findings[1]	Age at onset	Alive (A) or age at Death (D/ y)	Blood or plasma NH3[2]	Acid-base status[3]	Orotic aciduria[2]	Liver OTC[4] activity % of control[4]	Km (orn)[5] vs. control mM	Km (CP)[5] vs. control mM	CRM[6]	Gene defect	Reference
24.	b, i	?	D/8d	4.1		784	12					393
25.	a, b, c, d, e, f, g, h, i	4m	A	6			10					394
26.	a, b, d, f, g	14m	A	3.6			6					395
27.	a, b, d, e, h, i	5y	D/6y	7.6		220	10					396
28.	a, b, c, d, e, h, i	6m	A	3.3	MAl	43	14					397
29.	a, b, d, i	2y	D/10y	11		13	3.5			5.6 (trace mRNA)		398, 399, 400, 401
30.	a, b, d, f, g, h, i	9m	A	13		186						402
31.	a, b, i	3m	D/+2d	10		182	7.4					402
32.	a, b, d, j	9m	A	6		32	6.6					402
33.	b, c, g, h, i	3w	D/13m	3.3	MAc	4.3	5.8					402
34.	a, b, c, i	3y	A	7		550	7					403
35.	a, b, d, i	2y	D/11y	3.1		33	9.2	0.83/0.8	0.2/0.2	10		404
36.	a, b, i	2.5y	D/5.5y	1.8		38	8	0.75/0.8	0.2/0.2	10		404
37.	a, b, i	8m	A	4.5		69	22.5			48.5		404
38.	a, b, d, g, i	11m	A	7.3		704						227
39.	a, b, e, f, g	2m	D/2.3y	9.4			2					405
40.	a, b, d, f, g, h	2m	A	8.5	N	84	55					406
41.	b, c	32y	D/+5d	16		156	6.2	(valproate)				542
42.	a, b, d, e, h, i	4m	A	15								407
43.	b,c,d	10m	D/26m	4.6			8					408
44.	a,b,j	36y	A	1.8		6						409
45.	a,b,d,e,h,i	37y	A	7.3		2.3						409
46.	c,d	5y	D/7d	11.4	N		16	0.53/0.54	0.83/1.09			410
47.	a,b	2y	D/2.1y	2.3		206	4	0.59/0.54	0.87/1.09			410
48.	a,b,c,i	60h	D/17d	8.4		84	4					411
49.	a,b,d	12y	A	8.5	RAl	35						412

Table 7-1 (continued)

Case No.	Clinical findings[1]	Age at onset	Alive (A) or age at Death (D/ y)	Blood or plasma NH3[2]	Acid-base status[3]	Orotic aciduria[2]	Liver OTC[4] activity % of control[4]	Km (orn)[5] vs. control mM	Km (CP)[5] vs. control mM	CRM[6]	Gene defect	Reference
50.	a, b, g, i	7m	A	3								413
51.	b, d, e, g, i	12y	A	5.1		1360					R 141 ter	246
52.	c, h, i	6y	A	5.3			32	(valproate)				543
53.	a, b, c, f	2m	A	6		12	11					414
54.	a, b, d	19m	A				5				R 141 ter	247
55.	a, b, c, i, f	13m	A	9.6	MAl	9.3	7					415
56.	a, b, c, i, j	18m	D/+6d	5.9		96	12.5					416
57.	b, a	21y	D/+12d	5.7	MAl	42	5.5	(8d post-partum)				417
58.	a, b, h, i	22y	A	4				(3d post-partum)				417
59.	b, h, i	37y	A	5.9	MAl							417
60.	a, b, f, g, i	14m	A	7.7	N		1.2					418
61.	a	15m	A	3		64	10					419
62.	a, b, h	3.8y	D/+5d	3.5		5	15	(valproate)				420
63.	b	4.5y	A	11		8.8	7	0.28/0.4		20		421
64.	a, b, d, h, i	15m	A	8.6		2.7						422
65.	a, f, g, h	13m	D/6y	9			63	4.3/0.42	1.3/0.21			423
66.	a, b	30m	D/33m	10.5			3–16			3–16	Y 313 D	254
67.	a, f	8m	A	>23		>17	4					251
68.	a, b, c, d, e, g	2.3y	A	6.8	MAc	3.4	6					424
69.	a, c, d, e, g, i	22y	A	4.2		2.7		(valproate)				544
70.	a, d, f, g, j	6m	A	4.4		12.6						425
71.	a, d, f, g, j	5m	A	4.1			11.4					425
72.	a, b, d, e	13y	D			42.5	10.9					426
73.	a, b, d, e, g, i, j	2.5y	A	3.1		39						427
74.	b, d, f, g	3.2y	A	1.9		75	26					427

Case No.	Clinical findings[1]	Age at onset	Alive (A) or age at Death (D/ y)	Blood or plasma NH3[2]	Acid-base status[3]	Orotic aciduria[2]	Liver OTC[4] activity % of control[4]	Km (orn)[5] vs. control mM	Km (CP)[5] vs. control mM	CRM[6]	Gene defect	Reference
75.	b, f, g	4y	A	4.9		10.9						427
76.	b, c, d, f, g	1y	A	3.2								428
77.	a, b, d	6m	D	4.8, 25		109	13.5					429
78.	a, b, e, f, g, h	9m	A	20.5	MAc	32	10					430
79.	a, d, e, g	8m	A	5.2		19.8						430
80.	a, b, c, g, h	1w	A	4.4			10					430
81.	a, b, c, e, f, g, i	7m	A	7.1								430
82.	b, d, e	31y	A	7.2		8.1	(allopurinol)					431
83.	a, b, c, e	3y	D/7y				26					267
84.	a, b, c, d, e, j	4y	A	7.5	RAl	202		(central pontine myelinolysis)				432
85.	b, i	?	A				3.5	0.22/0.51	N		R 129 H	271
86.	a, b, e, h	3w	A	6.7			9.7	(orthotopic liver transplant)				433
87.	b, c, d, i	6.5y	A	3.4		2.7		(valproate at onset)				434
88.	b, c, g, j	3y	A	4.1		1.4						435
89.	a, b	3y	A	3.9	MAc		30	(APOLT)[7]				436
90.	a, b, g	19m	A	5.5		10.2	30	(APOLT)				437, 438
91.	b	3y	A	4.8			12.5				R 320 ter	439
92.	a, d, e, i	2m	A	5.9		4.8						440
93.	e, f	10m	A	1.5		24.9						440
94.	a,c,d	3m	A	3		9.4						440

Table 7-1 (continued)

Case No.	Clinical findings[1]	Age at onset	Alive (A) or age at Death (D/ y)	Blood or plasma NH3[2]	Acid-base status[3]	Orotic aciduria[2]	Liver OTC[4] activity % of control[4]	Km (orn)[5] vs. control mM	Km (CP)[5] vs. control mM	CRM[6]	Gene defect	Reference
95.	a, d, h	2m	A			2820	50				g→t, nt 663+1, int 6	274
96.	a	14m	A	4.7		613	100	0.25/0.63	0.48/0.34			275
97.	b, d	24y	D/+2d	7.6			32				del nt 892, 893	278
98.	a, b	16m	D/27m	8.5		18					T 44 I	279
99.	a, b, c, g, i	9m	A	9.9		42					R 320 ter	279
100.	a, b, d, g, h	5d	A	2.6	RAl	2.4					g→t, nt 78+1, int 1	441
101.	b, i	5y	D/+2d			8.2	14				L 148 F	442
102.	a, b, g	19m	A	5.5		110	<30	(APOLT)				443
103.	a, b, c, i	2y	D/25y	21.5	RAl	213		(post-partum)				444
104.	b, c	3d	D/12d	56	MAc	175	0 (shock liver)					445
105.	e, h	3y	A	3.9		90					del G, nt 664	293
106.	b, d, i	3.5y	A	5			23?				H 202 P	295
107.	a, b, c, d, e, g	5y	A	9.6		38		(valproate)				446
108.	a, b	29m	A	7.1		1.6	9.7	(APOLT)				438
109.	a, b, i	22m	A	0.8			22	(APOLT)				438, 541
110.	a, d, j	15m	D/+10d	9.2		1137	9.7					447
111.	b, c, j	52y	A	2.4		164	100?				No mutation in gene	448

Case No.	Clinical findings[1]	Age at onset	Alive (A) or age at Death (D/ y)	Blood or plasma NH3[2]	Acid-base status[3]	Orotic aciduria[2]	Liver OTC[4] activity % of control[4]	Km (orn)[5] vs. control mM	Km (CP)[5] vs. control mM	CRM[6]	Gene defect	Reference
112	b,h	11m	A	7.1	RAl	169						725
113	a,b,d,h	4y	D/+3d	6.5								726
114	a, b	28y	A	2.1	RAl	6.9		(bleeding gastric erosions)				727
115	b, d, i	53y	A	3.5		2.5					R60 ter	715
116	d, e, h	23y	D/+17d	7.4		888	19.4				N 161 S	719
117	a, b, e, h, i	12y	A/20y	13.4	RAl	70					G 195 R	720
118	a, e, i, j	40y	A/52y	5.5		23		(protein load; mother of 117)			G 195 R	720

1. a=vomiting, b=lethargy, coma, c=seizures, d=abnormal behavior, e=protein avoidance, f=delayed growth, g=delayed development, h= incoordination, i=positive family history, j=persistent CNS deficit.
2. Fold increase over upper limit of controls; values are those at onset.
3. N=normal, RAl=respiratory alkalosis, RAc=respiratory acidosis, MAc=metabolic acidosis.
4. OTC assayed at pH 7.7–8.3, as % of normal mean value by same method.
5. Km's measured at pH 7.7–8.3, over values for controls by same method.
6. Cross Reacting Material.
7. Auxiliary Partial Orthotopic Liver Transplant (APOLT).

Table 7-2. Clinical and laboratory findings in neonatal male probands with OTC deficiency, 1969–2002

Case No.	Clinical findings[1]	Age at onset	Alive (A) or age at Death (D/)	Blood or plasma NH_3[2]	Acid-base status[3]	Orotic aciduria[2]	Liver OTC[4] activity	Km (orn)[5] vs control	Km (CP)[5] vs control	pH optimum	CRM[6]	Gene defect	Reference
	Early onset group (0–28 days)												
1.	b, c, i	3d	D/5d	8	RA1		0.4						467, 468
2.	b, i	1d	D/2d				0						469
3.	b, i	?1d	D/?2d				0						469
4.	b, c, i	36h	D/9d	2	RA1		0.14	0.11/0.43	0.37/0.48	7.7–8.2			468
5.	b, c	36h	D/5d				0						470
6.	b, c, i	24h	D/5d	7			0.28						470
7.	a, b, c, i	40h	D/10d	5.3	MAc		2.3	55/0.52	0.31/0.57	9.5			471
8.	b, c, i	7d	D/25d	4.4		24	0						450
9.	a, b, c, e, i	15d	A	6	N	125	2.5			8.6–9.5			451
10.	a, b, i	6d	A	8	RAl		6	2.4/0.44	0.80/0.55	8.3			452
11.	b, d, i	24h	D/3d	2.7			0.3						472
12.	b, c, d, h, i	5h	D/54d	2.9			0.2						473
13.	b, c, i	24h	D/2.9d	3.2		34	0.6	2.3/6.6	10/4.16	8.4			474
14.	b, d, h, i	24h	D/2.5d	23			0						393
15.	d, i	24h	D/5m	1.2			0						394
16.	b, c, i	4d	A	7.4	RAl	30							475
17.	a, b, i	48h	A	17			0						476
18.	d, i	12h	D/24h		MAc		1.3						453
19.	a, b, i	2d	D/3d	26									477
20.	b, i	3d	A	12		75	0						478
21.	b, c, d, i	24h	D/6d	8.8			0						479

Case No.	Clinical findings[1]	Age at onset	Alive (A) or age at Death (D/)	Blood or plasma NH_3[2]	Acid-base status[3]	Orotic aciduria[2]	Liver OTC[4] activity	Km (orn)[5] vs control	Km (CP)[5] vs control	pH optimum	CRM[6]	Gene defect	Reference
22.	b, c, i	7d	D/28d	4.7			0.5				0		404
23.	b, c, d, e, i	24h	D/19d	4.2	MAc	138	0						228
24.	b, f, g, i	4d	D/2.5y	14.7								R 141 Q	230
25.	b, i	2d	D/8d	11.9		20	0					R 141 Q	230
26.	g, i	2d	A			41	2					R 277 W	242
27.	b, i	3d	D/3d			220							456
28.	a, b, c, i	3d	D/9d	26		5.6	0					P 225 L	237
29.	b, i	26d	A	12.4			1						480
30.	b, i	48h	A	4	RAl	21						gt->gc, nt 719, del intron 7	481
31.	b	4d	D/7d	>60			0					del nt 403	254
32.	b, i	72h	D/4d	>60			0					g->a nt 217 accept. spl. error	254
33.	b	5d	A	16.5			0	(liver trplt)				G 47 E	254
34.	b, c	24h	D/48h	>23		>17	1.4						251
35.	b, i	2d	D/10d	24			0					S 192 R	255
36.	b, g, i	5d	A	9.9		1120	7	20/0.7	N			D 196 V	255
37.	c, i	24h	D/10d	24		79	0					R 141 ter	256
38.	b, d, i	24h	D/48h	>20		>17	0	(same results, twin brothers)			del OTC gene	460	
39.	b	2d	D/3d	22.4	RAl		0					del nt 199, frame shift	482

Table 7-2 (continued)

Case No.	Clinical findings[1]	Age at onset	Alive (A) or age at Death (D/)	Blood or plasma NH_3[2]	Acid-base status[3]	Orotic aciduria[2]	Liver OTC[4] activity	Km (orn)[5] vs control	Km (CP)[5] vs control	pH optimum	CRM[6]	Gene defect	Reference
40.	a, b	24h	A	6.7	RAl	58	0	(orthotopic liver transplant)					433
41.	b, i	24h	D/6d				0				0	Donor splice error, del exon 5	273
42.	b, c	4d	D/5d				0					T178 M	267
43.	a, b	7d					20					R 277 W	267
44.	b, c, g, i	10d	A	5.8		1.7	2			9.5		G 269 E	269
45.	b, i	2d	D/4d				0.3					Y 167 ter	271
46.	b, i	4d	A				1.3	(jejunal biop.)		7.7		R 129 H	272
47.	b, i	3d	D/10m	15.8		82.6						H 214 Y	279
48.	a, b, i	3d	D/6d									R 141 Q	279
49.	b, i	3d	D/8d				0					a→c, +4 int 1	277
50.	b	1d	D/15d				0					g→t, 867+1 int 8	274
51.	b	2d	D/3d				0					g→c, 540+1 int 5	274
52.	b, h, i	30h	D/3d	24									461
53.	b, d, i	2d	D/6d	15.6	MAc		0					H 302 Y	267
54.	b, i	1d	D/1m	13.3			0					R 141 ter	267
55.	b, i	2d	A	26		46		(liver trplt at 7 m)				R 141 Q	252, 459
56.	b, c, g, i	2d	D/6m	14		927							441
57.	b, i	3d	D/5d				0					P 225 R	483
58.	b, i	3d	D/5d				0.15	1.5/0.4	1.33/0.5			P 225 L	483

Case No.	Clinical findings[1]	Age at onset	Alive (A) or age at Death (D/)	Blood or plasma NH_3[2]	Acid-base status[3]	Orotic aciduria[2]	Liver OTC[4] activity	Km (orn)[5] vs control	Km (CP)[5] vs control	pH optimum	CRM[6]	Gene defect	Reference
59.	a, b	3d	A	11.7								M 205 T	279
60.	b, e	2d	A	10.3		58	<20	(orthotopic liver transplant)					484
61.	b, c, g	2d?	A	3.8		53	<20	"					484
62.	b, e, i	2d	A	19		82	<20	"					484
63.	g, i	1d	A	3			0	(treated at birth)				E 181 G	462
64.	c, d, g	3d	A	45			1						485
65.	a, b	3d	D/7d	39									485
66.	b, i	2d	D/10d	31		177	>5					M 206 R	464
67.	b, c	53h	D/1.2y	20			0					297del gact	728

1. a =vomiting, b=lethargy, coma, c=seizures, d=abnormal behavior, e=protein avoidance, f=delayed growth, g=delayed development, h= incoordination, i=positive family history, j=persistent CNS deficit.
2. Fold increase over upper limit of controls; values are those at onset.
3. N=normal, RAl=respiratory alkalosis, RAc=respiratory acidosis, MAc=metabolic acidosis.
4. OTC assayed at pH 7.7–8.3, as % of normal mean value by same method.
5. Km's measured at pH 7.7–8.3, over values for controls by same method.
6. CRM=cross-reacting material by various immunoassays, as % of control livers.

Table 7-3. Clinical and laboratory findings in late-onset male probands with OTC deficiency, 1969–2002

Case No.	Clinical findings[1]	Age at onset	Alive (A) or age at death (D/)	Blood or plasma NH_3[2]	Acid-base status[3]	Orotic aciduria[2]	Liver OTC[4] activity	Km(orn)[5] vs. control	Km(CP)[5] vs. control	pH optimum	CRM[6]	Gene defect	Reference
1.	a, b, c, f, g, h, i	6m	D/14y	4	N		74	0.89/1.26	0.34/1.45				371, 402, 486, 487
2.	a, b, f	35d	D/4m		MAc	10	4						488
3.	a, b, c, i	8y	D/+7d	9.1	RAl		9	0.44/0.46	0.31/0.54	7.5			449, 452
4.	a, b, c, i	4y	D?				6						489
5.	a, b, c, d, i	3m	D/3y	6			4	9.7/0.6	0.8/0.3	8.5			490
6.	a, b, g	9m	A	2		341	5						491
7.	a, b, e, d, g	6m	D/13m	4.7			1.6	3.9/0.41	1.14/0.41	8.8	8		493
8.	b, d	11m	D/?	3	N	140							494
9.	a, b, c,	4m	D/7m	10.4		21	20						495
10.	b, d, h	4y	D/+14d	3	MAc	580	10.9						496
11.	a, b, i	13m	A	3.8	RAl	750	19	8.3/0.8	N	8.5		R 277 W	248, 497
12.	a, b, d	11y	D/+24d	9		150	2.8	(valproate)					545
13.	a, b, i	13m	A	2.1	N	1070	16	9.8/0.8	N	8.5		R 277 W	248, 497
14.	a, b, c, i	9m	D/3.5y	1.5	N	720	3.4	0.79/0.40	0.65/0.52	7.7			498
15.	a, i	2m	A	2.6	(benzoate from birth)		2						477
16.	a, b, c, e, i	6m	A	3.9	N	7.5	10	1.82/1.48	0.015/0.24		75		499
17.	a, b, c	2ly	D/+6d	4.8	N	17	8.4	1.5/0.4	1.1/0.4	8–8.3			500
18.	a, b, d	12y	D/+1d	7		579	4						402
19.	a, b, d, i	8m	D/12m	17.5		140	4.5						402
20.	a, b, f, i	5y	A	3.8	N	76	16						501
21.	a, d	23y	A	3.2		20	26						502
22.	a, b, i	12y	D/12y	12		267	1.6						455
23.	a, b	6m	A	4.6		1180	10			9.5			503
24.	b, d	12y	A	5.8		525							504
25.	a, b, c, h, i	10y	D/10y	34		130							505

Case No.	Clinical findings[1]	Age at onset	Alive (A) or age at death (D/)	Blood or plasma NH_3[2]	Acid-base status[3]	Orotic aciduria[2]	Liver OTC[4] activity	Km(orn)[5] vs. control	Km(CP)[5] vs. control	pH optimum	CRM[6]	Gene defect	Reference
26.	b, d, e, g, i, j	30y	A	4.2		58.5	0.94					M 56 T	463, 506
27.	a, b, h, i	17m	A	6.6		15	1.7						504
28.	a, b, d, g	12y	D/12y	5.7		579	3.2						504
29.	a, b, d	8m	D/12m	14		140	3.6						504
30.	a, b, d	10m	A	6		848	5.8						504
31.	a, b	9m	A	17	RAl	809	5.7						412
32.	a, b, i	6m	A	3.2		8900	2.3						413, 507
33.	b, i	15m	D/+3d	5.2		143	54	3.8/0.33	0.42/0.29	9		R 277 Q	508
34.	a, b	10y	D/+6d	2.9			6.2	(valproate)					546
35.	a, b, i	14y	D/+14d	7		103	10	0.5/0.4	0.2/0.4	7.7	10		509
36.	a, b, c, d, i	56y	D/56y	51		442	1.3				trace	R 40 H	458
37.	a, b, d	46y	D/46y	15.5		423	2.6				trace	Y 55 D	458, 510
38.	a, b, c, d, i	17y	D/17y	50.5		451	5.5				5–10	R 40 H	458
39.	a, d	9y	D/19y	3.4		511	7						511
40.	a, b, d	9m	A	19		1330	5.7						511
41.	a, f, e	6m	A	4		70							512
42.	a, b	15y	D/+3d	5.4			10						513, 515
43.	c, g	10m	D/18m	3.6		1.6	50	2.0/0.33	N			R 277 Q	514
44.	a, b	8y	A	3			9.8	8/0.7	0.07/0.12			R 277 W	248, 255
45.	a, b, d	7y	D/+1d	15.6	MAc	471	0.8					R 94 T	254, 261
46.	b, d, g, i	7m	A	8			3.2				0	L 304 F	254, 261
47.	a, b, c, d, f, i	3.5m	D/7.5m	>23		>17							251
48.	a, b, h, i	30y	A	5.5		23							516
49.	d, g, i, j	8m	A				5.5–20					R 129 L, donor splice error, ex 4	259
50.	a, b, c, i	6m	D/5y	3.2			6.7					M 268 T	255
51.	a, b, c, f	4y	A	3.5			22					T 264 A	255

Table 7-3 (continued)

Case No.	Clinical findings[1]	Age at onset	Alive (A) or age at death (D/)	Blood or plasma NH_3[2]	Acid-base status[3]	Orotic aciduria[2]	Liver OTC[4] activity	Km(orn)[5] vs. control	Km(CP)[5] vs. control	pH optimum	CRM[6]	Gene defect	Reference
52.	a, b, c, e	38y	D/+3d	5.9	N		5.5						517
53.	b	15m	D/+3d				60	3/0.22	N			R 277 Q	266
54.	c	10m	D/18m				50	2/0.22	N			R 277 Q	266
55.	a, d.i	4y	A	1.4		3.6	18						518
56.	b	2y	A	4			2.4					delE, ΔG AA, 309, 310	264
57.	b	8m	A	3								R 277 Q	264
58.	b	8y	A	7.6								P 225 T	264
59.	a, b, d	9.5y	A	5.6	MAl	318	2.6						519
60.	b, c	4y	A				6				trace	L 304 F	273
61.	a, b, i	5m	A			526	5					?	273
62.	b, i	14y	D/18y				3					R 40 C	267
63.	b, i	9y	D/+2d				7.5					R 40 H	267
64.	i	13y	A				25					R 40 H	267
65.	a, b, c	6m	D/+6d				6					R 277 W	267
66.	b	1y	A	4		78	14					T 343 L	282
67.	a, b, i	1y	D/2.5y	23.5			35					P 220 A	282
68.	b	2.5y	D/+6d				19					Y 176 C	282
69.	a, b, i	4y	A	4.3		89	23					L 88 N	282
70.	a, b, i	5m	A			526	<5					g→a, 540 +1, int 2	274
71.	b, e	10m	A				5					In frame del E272/273	283
72.	a, b, c	9y	D/+5d	17.6			12					R 40 H	280
73.	a, c, i	15y	D/+9d	31.7			3.4					"	280
74.	b	17y	D/+7d	46.4			2.4					"	280

Case No.	Clinical findings[1]	Age at onset	Alive (A) or age at death (D/)	Blood or plasma NH_3[2]	Acid-base status[3]	Orotic aciduria[2]	Liver OTC[4] activity	Km(orn)[5] vs. control	Km(CP)[5] vs. control	pH optimum	CRM[6]	Gene defect	Reference
75.	a, b, c, h	13.4y	D/+5d	8.2		836	1.8					F 322 C	520
76.	b, d, i	12y	D/14y				3.3					A 208 T	284
77.	b	11y	A	4.2			5			7.7		R 40 H	286
78.	a, b	17y	A	2.7			7	N	N	7.7		R 40 H	286
79.	b, c, d, i	6y	D/+6d			304	3.6					A 208 T	287
80	b, i	12y	D/+2d	60		446						R 40 C	718
81	b, i	14y	D/+4y	4.8		52	7.2	(brother of 80)				R 40 C	718
82	a, b	26y	D/+1d				10.6	(liver trplted → death of recipient				R 40 H	

1. a=vomiting, b=lethargy, coma, c=seizures, d=abnormal behavior, e=protein avoidance, f=delayed growth, g=delayed development, h=incoordination, i=positive family history, j=persistent CNS deficit.
2. Fold increase over upper limit of controls; values are those at onset.
3. N=normal, RAl=respiratory alkalosis, RAc=respiratory acidosis, MAc=metabolic acidosis.
4. OTC assayed at pH 7.7–8.3, as % of normal mean value by same method.
5. Km's measured at pH 7.7–8.3, over values for controls by same method.
6. CRM=cross-reacting material by various immunoassays, as % of control livers.

deficiency. The OTC activity in the first case was later shown to be 5% of normal on liver biopsy (Table 7-1). Subsequently the second girl died and autopsy OTC activity was 7% of an autopsy control[370]. The identical twin of the second girl developed symptoms at age 9 but no OTC assay was ever performed[369,370]. She is included as the third case in Table 7-1. Further studies on these three girls were published in 1968 and 1969[371,372]. From 1962 to 2002, we located a total of 118 female probands whose clinical data were adequate to be reported in Table 7-1. Six patients (numbered 41, 52, 62, 69, 87 and 107) had the onset of symptoms precipitated by treatment with valproic acid or sodium valproate, an anti-seizure medication. These patients will be discussed at the end of this chapter. Their symptoms, age of onset, outcome, plasma ammonia and urinary orotic acid excretion and acid-base status were not included in the calculations reported in Table 7-5 because the drug may have affected these variables. The liver OTC activities were included in our calculations because valproate was not known to affect enzyme activity.

A key characteristic of these affected females is the later time of onset of symptoms compared to that of the neonatal hemizygous males (average 5.9 years vs. 2.7 days, Table 7-5). In this X-linked trait the females have a normal and an abnormal allele and the random inactivation of their X-chromosome determines how much of the abnormal gene will be expressed (Lyon hypothesis)[2]. Ninety five percent of *symptomatic* females had 12–20% of normal mean liver OTC activity whereas 95% of neonatal onset males had 0.2–2% of normal activity (Table 7-5).

Among the 112 non-valproate female patients reported, only 12 had a neonatal onset (within 28 days). They are patients numbered 4, 7, 15, 19, 20, 21, 33, 48, 80, 82, 100 and 104 in Table 7-1. The liver OTC activities in the 9 cases with data available ranged from 0–33%. Number 104 with zero activity does not reflect extreme lyonization but is due to severe post-partum hypoxia, hypotension and ischemic necrosis of the liver; in this case the carbamyl phosphate synthetase activity was only 25% of normal. The average OTC of the other 9 cases is 12.8 ± 10.1%(SD) of normal, compared to the average of 16 ± 17% of all 77 female cases with liver assays (Table 7-5). Other than case 104, only cases 20, 33 and 48 died, at ages 3.9y, 13m and 17d respectively. The mortality of 33% in the early onset females (4/12)is only slightly higher than that of the 34/118 females (29%) overall.

Early onset did not correlate with lower OTC activities (r = −0.03) but may have reflected the protein load that was fed to these infants, usually cow's milk formulas. Human breast milk, which is lower in protein, seemed to be better tolerated.

Most of the 118 female patients had an onset after 28 days (average 5.9 ± 11 years, Table 7-5) and 13 did not become ill until over age 21 (range 22–53), (Table 7-1). Seventy-one percent presented with vomiting, 81% with lethargy/coma, only 28% with seizures and 42% with abnormal behavior (Table 7-4). Protein avoidance was mentioned in 31% of these females. Unfortunately delayed growth (21%) or delayed mental development (34%) was often present when the diagnosis was recognized. Ataxia and incoordination were reported in 25% and a persistent CNS defect in 10%. A positive family history was obtained in 44% of these patients (Table 7-4). The age at death was recorded in 34 patients and occurred at a mean of 7.0 ± 8.4 years (Table 7-5).

The elevations of blood or plasma ammonia over the upper limits of normal ranged from 0.8- to 56-fold; the mean value was 6.8 ± 6.3-fold (Table 7-5) and the median value was a 5.7-fold elevation. Acid-base status was normal in 11 of the 31 patients where data were adequate to judge. Nine reported a respiratory alkalosis, six a metabolic acidosis, and five a metabolic alkalosis (Table 7-5). This variation is evidence that the acid-base status is not diagnostically useful in these

Table 7-4. Clinical symptoms in neonatal-onset males, late-onset males and female patients with OTC deficiency, in percent of total cases

Symptoms	Neonatal-onset males	Late-onset males	Females
a. vomiting	16	68	71
b. lethargy, coma	91	87	81
c. seizures	33	28	28
d. abnormal behavior	16	30	42
e. protein intolerance	6	7	31
f. delayed growth	2	6	21
g. delayed development	13	9	34
h. incoordination	5	7	25
i. positive family history	78	46	44
j. persistent CNS deficit	0	4	10
Total cases	67	82	118

Table 7-5. Baseline and laboratory data in neonatal-onset males, late-onset males and female patients with OTC deficiency

	A. Neonatal males		B. Late-onset males		C. Females	
	mean +/- SD (n) (95% confidence limits)					
1. Age at onset	2.7±2.4 (2.1–3.3)	(67) days [b]vs. B, C	8.0±10.5 (5.8–10.3)	(82) years [a]vs. C	5.9±11 (3.8–7.9)	(115) years
2. Age at death	43±145 (-1.1–87.1)	(45) days [b]vs. B, C	11.2±11.7 (7.7–14.7)	(46) years [a]vs. C	7.0±8.4 (4.1–10)	(34) years
3. Blood, plasma NH_3	15.2±13.8 (11.2–19.2)	(48) times nl. [a]vs.B, [b]vs.C	10.4±12.8 (7.3–13.6)	(66) times nl.	6.8±6.3 (5.6–8.0)	(113) times nl.
4. Orotic aciduria	144±278 (26.5–261)	(24) times nl. [a]vs.B	526±1280 (154–898)	(49) times nl.	176±400 (85–267)	(78) times nl.
5. OTC activity	1.1±3.1 (0.24–2.0)	(52) % control [b]vs.B, C	11.6±14.5 (8.2–15)	(74) % control	16.1±17.2 (12.2–20.1)	(77) % control
6. Acid-base status		(n=12)		(n=15)		(n=31)
resp. alkalosis	58%		20%		29%	
metab. alkalosis	0%		7%		16%	
metab. acidosis	33%		20%		19%	
normal	8%		53%		35%	

a. $p<0.05$ by 2-sample t-test.
b. $p<0.001$.

female patients. Moreover, lactic acidosis causing metabolic acidosis in moribund patients suggests erroneously a secondary cause of hyperammonemia by an organic acidosis such as propionic aciduria. Urinary orotic acid levels were reported in 78 patients with an average elevation of 176 ± 400-fold over the upper normal values, the distribution being skewed markedly toward higher values. Thus urinary orotate was a much more sensitive indicator of a block in ammonia removal than blood/plasma ammonia levels (Table 7-5).

Liver OTC activity averaged 16 ± 17% of normal in 77 patients where measured and varied between 0.6 and 69 percent of controls as one would expect from random inactivation of one X-chromosome. Case number 9[380] died at the age of 10 months with 69 percent of control OTC activity assayed at pH 8.3 in the liver obtained at autopsy, a value within the normal range. The enzyme in case 9, however, was abnormal in that the K_m for carbamyl phosphate was four times higher than the control whereas the K_m for ornithine was normal. The maximal velocity (V_{max}), varying carbamyl phosphate, was 84% of a control biopsy of liver. These kinetics are puzzling because in an OTC heterozygote the clones of liver cells carrying the normal X-chromosome should account for most of the measured activity[3]. If the mutant OTC is also partly active, its kinetics will be obscured by mixing with normal OTC in the biopsy sample. The same problem exists with the small changes in K_m seen in other female patients in Table 7-1. Meaningful kinetics of a mutant OTC can only be derived from studying the hemizygote male liver enzyme, where only the abnormal gene is being expressed.

We calculated a correlation coefficient by linear regression between OTC activity (% normal mean) as the independent variable (x) and time of onset of symptoms (years) as the dependent variable (y) for 77 paired samples from Table 7-1. The correlation coefficient (r) had a value of 0.47, with a t-value of 4.5 and a p-value of 2.7×10^{-5}; the low p-value is due to the high number of 73 pairs. The slope of the linear regression was 0.193 and the intercept was −0.270 years on the y-axis (at 0% OTC activity). The residual plot showed a negative deviation, which suggests that a linear regression was not a good fit. Nevertheless there is a statistical correlation whereby the measured OTC activity in liver does roughly predict the time of onset of symptoms but there are many obvious exceptions when Table 7-1 is examined.

We reviewed all our detected cases of OTC deficiency and looked

Table 7-6. Confirmation testing of female carrier status for OTC deficiency

Case No.	OTC activity (% control)[a]	Loading, Challenge Tests: NH$_4$Cl	Loading, Challenge Tests: Protein	Loading, Challenge Tests: Allopurinol	RFLPs (endonuclease employed)	Gene sequence	References
1		+pNH3[b]					372
2	25(intestine)	+pNH3					372
3	40	+pNH3					374
4	43		+pNH3				382
5	40	+pNH3					384
6	69	+pNH3					384
7	97	?+pNH3					384
8	106	-pNH3					384
9	131	-pNH3					384
10	20		+pNH3				449
11		+pNH3	+OA[c]				450
12		"	"				450
13			"				450
14	45(intestine)		+OA				451
15			+glutamine[d]				387
16	13		+pNH3				452
17	46						452
18	50		−pNH3, -OA				397
19	35						396
20			+pNH3, OA				453
21			+pNH3, OA		MspI		454
22			+pNH3, OA		MspI		455
23			+pNH3, OA		MspI		407
24			+pNH3, OA		MspI		413
25			+pNH3, OA		MspI		413

Case No.	OTC activity (% control)[a]	Loading, Challenge Tests			RFLPs (endonuclease employed)	Gene sequence	References
		NH_4Cl	Protein	Allopurinol			
26			+OA		PstI, MspI, TaqI		456
27			+OA		MspI		567
28			+OA				458
29			+OA				458
30			+OA				458
31			+pNH3, OA				458
32						R141Q	459
33						R141Q	459
34				+OA		R129H	271
35			+OA			del gene	460
36			+OA			del gene	460
37					TaqI, EcoRI		461
38				+OA			441
39				+OA			438
40				+OA		H202P	295
41						E181G	462
42						M56T	463
43				-OA	(?gonadal mosaicism)	M206R, in sons	464

a. % of mean control values.
b. Plasma ammonia elevation (+) or no elevation (-).
c. Urinary orotic acid elevation (+) or no elevation (-).
d. Plasma glutamine elevation.

for data concerning the mothers, grandmothers, sisters, aunts and cousins of deficient males and females who as relatives did not qualify to be included in Table 7-1 because they were not probands. Most were not symptomatic unless challenged but had their OTC deficiency carrier status confirmed by liver enzyme assays, challenge tests or gene studies. These 43 cases are collected in Table 7-6. All were female relatives of proven probands. Eight cases gave a history of protein intolerance (cases 1–3, 5, 6, 10, 12, 20) and 32 showed abnormal results on loading tests with ammonium chloride, protein or allopurinol. Only cases 2 and 3 had elevated fasting plasma ammonia levels and only case 2 had an elevated fasting plasma glutamine. OTC assays in liver (rarely in intestinal) biopsies were done in 14 cases from 1968 to 1980. After that time these invasive procedures were supplanted by challenge tests or gene analyses (Table 7-6). Ammonium chloride challenge tests were replaced by a single meal of mixed protein (usually 1 g/kg). Urinary orotic acid measurements, usually expressed as μg/mg creatinine, began to supplant plasma ammonia levels in protein challenge tests because they were thought to be more sensitive. In 1986 DNA analysis using restriction fragment length polymorphisms (RFLPs) began to be used to detect carriers of OTC deficiency. By 1996 sequencing of all 10 exons and intron borders was applied to carrier detection (Table 7-6). The oral administration of 3g of allopurinol followed by four 6-hour collections or one 24-hour collection of urine for orotate and/or orotidine determinations replaced protein loading tests because side-effects were less even though sensitivity and accuracy were no greater (Chapter 8).

Case number six in Table 7-6, the aunt of a proband, had life-long protein intolerance and her OTC activity on needle biopsy was 69% of the control mean value, close to the lower limit of normal which is 76% by this method[384]. The value in the patient was 22.7 units per mg protein whereas two standard deviations below the normal mean was 25.1. Thus it appears from such maternal carriers as cases 3–6 and 18 that OTC activities in the range of 40–69% of normal are in little excess functionally within the mitochondria *in vivo* in spite of OTC's high *in vitro* activity compared with carbamyl phosphate synthetase and argininosuccinate synthetase, either of which have been considered to be rate limiting in the urea/ornithine cycle. Considering that the pH optimum of OTC is 7.7 whereas the pH in the mitochondrial matrix is 7.4, and that the substrate is the form of ornithine with an uncharged delta amino group (pK 8.7), which is only 5% of total ornithine at pH 7.4, then OTC activity in the mitochondria functions from moment to moment at a rate controlled primarily by the mitochondrial concentration of ornithine, which seems to be channeled to OTC by the ornithine transporter(154). This may be why an acute protein or ammonium load causes symptoms or elicits abnormal plasma ammonia or urinary orotate responses in these usually asymptomatic female carriers. While mildly or severely symptomatic heterozygotes often develop various CNS defects(406,466), Maestri et al. found that "asymptomatic" carriers restricted their intake of meat and dairy products and excreted less urea and total nitrogen than control women(466). Batshaw et al.(407,466) have suggested that heterozygotes who develop hyperammonia after protein loading are at risk for chronic cerebral damage and for episodes of fatal spontaneous hyperammonia; they should be considered for alternate pathway (phenylbutyrate) therapy.

The liver OTC activity in female case 63 in Table 7-1 is puzzling(421). The female liver contained 20% of normal OTC protein (36kD) and ~20% of normal OTC mRNA but the activity of the OTC was only 7% of normal at pH 8 even though it showed normal K_ms for both substrates and no apparent shift in pH optimum. The gene defect was not defined. It is possible that active normal and inactive mutant monomers were randomly forming heterotrimers in the mitochondrial matrix with only one third containing active monomers. Such mixtures have been achieved in CHO cells transfected with 141Q-containing plasmids mixed with normal 141R-containing plasmids(454). R141Q protein has no activity but is 36kD in size. One or two of the catalytic sites in these heterotrimers remains active and no dominant negative effect was seen(454).

OTC Deficiency in Males

Tables 7-2 and 7-3 summarize the 149 clinically described cases we have found of *male probands* with OTC deficiency reported between 1969 and 2002. They are divided into a neonatal group of 67 with onset of symptoms from birth to 28 days and a delayed onset group of 82 with onset from 29 days on to the oldest reported male who was

56 years of age. Two late onset cases (numbered 12 and 34) whose symptoms were precipitated by valproate treatment are not included in the calculations except for liver OTC activities.

The onset of first neonatal symptoms usually occurred within 24 to 48 hours of birth and when described as beginning at 3 to 6 days, symptoms probably were not recognized as those of hyperammonemia. In the neonatal group only 16% manifested vomiting which was found in 68% of the late onset group (Table 7-4). The newborns presented with lethargy and/or coma in 91% compared with 87% in the late onset group. Seizures were mentioned in 33% of neonatal and 28% of late onset cases respectively. Abnormal behavior was described in 16% of neonatal and 30% of late onset males. In newborns hypertonia, hypotonia, apnea or hyperventilation often occurred, but these are non-specific signs of many neonatal problems. Protein intolerance in 6% of neonates was implied by refusal to ingest or vomiting of a formula, usually based on cow's milk. Only 7% of late onset males had protein intolerance. Delayed growth occurred in 2% and 6% and delayed development in 13% of neonatal and 9% of late onset respectively. Incoordination, usually ataxia, was noted in 5% and 7% respectively of the two groups. A persistent CNS deficit was mentioned in 4% of late onset cases only. Seventy-eight percent of neonates had a family history consistent with OTC deficiency but only 46% of late onset males had such (Table 7-4). The incidence of any of these symptoms is only approximate because the case reports vary widely in the details of their clinical synopsis. A series studied prospectively gathers more complete and reliable data.

In Table 7-5 we compare the age at onset of symptoms and age at death in those who died among neonatal and late onset males and females. The neonatal males became symptomatic at 2.7 ± 2.4 days, markedly sooner than the late onset males, 8.0±10.5 years, and the symptomatic females, 5.9 ± 11 years. The 95% confidence limits of the onset times were 2.1–3.3 days, 5.8–10.3 years and 3.8–7.9 years respectively. The mortality rate of neonatal males was 69% and on average the 46/67 neonatal males who died only lived 39 days; the survivors were listed as alive at the time of the report but we suspect that most died by 1–4 years in spite of alternate nitrogen pathway therapy (benzoate, phenylbutyrate). Maestri et al.[(522)] in the large Johns Hopkins series reported that 32 neonates died in the initial episode of coma and had a median survival of 144 hours, with none lasting more than 19 days; among the 40 infants who survived the initial hyperammonemic episode the median survival was 3.8 years and only 9 were alive at the time of the report. At the end of 2002 only liver transplantation secured long–term survival in early onset males. In Table 7-3 the late onset male mortality was 55% and in the 46 who were reported as dead among the 82 case reports death occurred on average 3.2 years after onset, which testifies to the fragile state of many of these patients in spite of our best therapy. The reported mortality in our females was 34/118 (29%).

Blood/plasma ammonia levels at *onset* were 15.2 ± 13.8 times higher than the upper limit of normal in neonatal males compared with a value of a 10.4 ± 12.8-fold elevation in late onset males ($p<0.05$) (Table 7-5). *Peak* ammonia values in neonatal males have been reported by others to be much higher than in late onset males and these high ammonia levels correlate with brain damage and early death[522]. Urinary orotic acid excretion per mg creatinine surprisingly was lower in neonates than in late onset males ($p<0.05$) (Table 7-5). We have no explanation for this anomaly. As expected, liver OTC activities were much lower in neonatal onset (1.1 ± 3.1% of normal) than in late onset males (11.6 ± 14.5%) and the 95% confidence limits, 0.24–2.0% vs. 8.2–15% respectively did not overlap.

We calculated correlation coefficients (r) for *neonates* between liver OTC activities (% normal mean) and time of onset (days after birth) from Table 7-2. With 51 matched pairs the r-value was 0.387 with a t-value of 2.939 and a p-value of 0.005. The slope of the linear regression was 0.329 and the intercept on the time axis was 2.24 days at 0% activity. When we did the same analysis on late onset males (Table 7-3) the 69 pairs gave an r-value of −0.210 with a t-value of −1.76 and a p-value of 0.08. The slope was −0.136 and the time of onset intercept was 9.3 years at 0% OTC activity and 69% OTC activity at zero years of onset. These late-onset negative correlations are puzzling because we expected a later date of onset as the liver activity increases. Inspection of the paired values shows this was not true in many cases. For example in cases 1, 9, 21, 33, 43, 51, 53, 54, 67, and 69 the activities ranged from 20–74% but the onset ranged from 0.3 to 4 years. We saw the same kind of discrepancy in some of the symptomatic females and have no explanation for this phenomenon.

The presence of respiratory alkalosis at onset in neonates (55%) is due to hyperventilation caused by the stimulation of the respiratory center by high plasma ammonia concentrations. It is often masked by a coexisting metabolic acidosis, usually caused by lactic acidosis.

Since 1979 a number of the neonatal survivors and severely OTC deficient late-onset patients have been maintained with the current therapy of low protein diets, sodium phenylbutyrate, carnitine and citrulline or arginine supplements. Still they remain prone to acute hyperammonemic episodes, brain damage or death[511,522]. Late-onset cases with liver activities above 10% can often be managed by the above therapies. Until gene therapy becomes possible, only total liver transplantation or auxiliary partial orthotopic liver transplantation (APOLT) gives long term survival but immune-suppression and citrulline therapy are then necessary (see neonatal cases 33, 40, 55, 60–62 and female cases 86, 89, 90, 102, 108 and 109).

Orotic acid excretion in the urine has been measured by many methods. The initial ones were relatively nonspecific. One of the standard colorimetric methods[523] measures both orotic acid and orotidine. New methods can separate orotate from orotidine[241]. Whatever the method employed it is apparent from Table 7-5 that the fold-elevation over the upper normal orotic acid level in the urine is more sensitive in confirming the diagnosis than the increase in blood or plasma ammonia or even plasma glutamine as noted below (Table 7-7). A single specimen of urine collected during an episode of hyperammonemia can be assayed for orotic acid and creatinine, expressed as micromoles of orotic acid per millimole of creatinine and allows a rapid diagnosis of a block in disposition of carbamyl phosphate in mitochondria.

The liver OTC activity in both male and female patients (Tables 7-1, 7-2, 7-3) was measured by a number of methods. Originally many of the studies used the method of Brown and Cohen[539] which employs glycylglycine buffer at pH 8.3, 10 mM ornithine and carbamyl phosphate and extraction of OTC with cetyltrimethylammonium bromide, a cationic detergent that agglutinates mitochondria and releases their enzymes. Some authors employed Tris buffer in spite of evidence that Tris inhibits OTC activity[98,141] and the inhibition becomes stronger as the pH rises. Glycylglycine reacts with carbamyl phosphate to form a colored product that increases the background color and requires a blank that corrects for this. Employment of buffers that are noninhibitory such as triethanolamine (pK_a 7.7) and establishment of the pH optimum of human liver OTC as pH 7.7[89] has made possible reproducible values among investigators[525,540]. For greater sensitivity an OTC assay using ^{14}C-radio-labeled ornithine, HPLC separation of citrulline and automated counting of radio-labeled citrulline has been reported[526].

Among the male *neonatal* group, as expected, the OTC activities are very low and in many cases essentially zero, accounting for the early death of these patients before any adequate therapy was available. A remarkable case is number 10[452] who became ill at six days but survived and had six percent of the control values in the liver. The mutant OTC had a shift in pH optimum from 7.7 to 8.3 and a six-fold increase in Km for ornithine at pH 8.3. At pH 7.4, using saturating 5 mM ornithine and carbamyl phosphate concentrations, the mutant OTC activity was only 2% of the control OTC. These kinetic defects should impair OTC activity severely in the mitochondria where the pH is 7.4 and ornithine varies from 0.3 to 1 mM[152]. This case resembles that of the spf mutant mouse (Chapter 6) which is due to an H117N change. Neonatal case 7[471] is also a pH mutant but the K_m for ornithine is 100-fold higher than normal. The group ionizing at the active site during catalysis (pK~6.6) in normal OTC[89] is shifted to a pK~9.5 in this mutant. Cathelineau[527] speculated that this is due to substitution of a lysine for the histidine at position 117 in pOTC but no gene studies have been done in this family.

The P225L mutation of pOTC in neonatal case 58 caused an activity almost zero but still allowed K_m studies that showed 3–4-fold K_m increases for both substrates. In case 28 this P225L mutant had zero activity as did the P225R mutation in case 57, but the P225T mutation allowed a late onset (OTC not measured) in case 58 in Table 7-3. At position 225 L,V and F as well as P are found in OTCs of other species[172,173] but L, R and T are not tolerated in human OTC. The reason why a proline is essential at position 225 in pOTC (193 in the OTC monomer) in the loop between the $\beta 7$ strand and the $\alpha 7$ helix in the human OTC monomer is not apparent from our analysis.

Late onset male cases 11, 13 and 33 are also pH mutants but here we know the gene defect: in 11 and 13 it is R277W and in 33 is R277Q (Table 7-3). They all show a 10-fold increase in the K_m for ornithine and a shift of pH optimum to 8.5–9. Our analysis in Table 5-5 indicates that these amino acid changes break an essential salt-bridge to D196 and impair closure of the domain where the substrates bind (Chapter 5). Table 7-3 cases 43,53 and 54 also had the R277Q defect and case 44 had the R277W error: they too showed the ~10-fold increase in the K_m for ornithine and normal K_m for carbamyl phosphate. The OTC activities in the R277W mutant were lower (9.8–19%) than the R277Q mutants (50–60%); the reasons are unclear.

An OTC liver activity measured at the usual pH optimum and saturating substrate concentrations may not correlate with the amount of functional OTC protein unless the K_ms and pH optimum of the enzyme are also measured. An example is case 7 in the late-onset males[(493)]. The OTC activity at pH 7.7 was 1.6% of controls but a later immuno-assay showed that the liver contained 8% of normal OTC protein. The explanation is the shift in pH optimum to 8.8 and the K_m of 3.9 mM rather than the normal value of 0.4 mM. When the assay was done at pH 8.8 with a saturating and not inhibiting ornithine concentration of 50mM, the activity was ~8% of normal. Kamoun and Rabier[(528)] have suggested that pH mutants can be suspected if the OTC assays are done at substrate values above and below the normal K_ms.

The late onset male group shows a wide spectrum of activities but one type of patient is difficult to understand. This is a patient who survives for months to many years with a low measured liver OTC activity and a history that gives no clue to previous hyperammonemic symptoms. The most extreme example of this phenomenon is patient number 36[(458)], who was 56 years of age at onset of symptoms and had only 1.3 percent of normal activity measured at pH 8.5 in 50 mM ornithine (the normal pH optimum is 7.7 and 50 mM is inhibitory). Immunoblotting of OTC protein showed only trace amounts. The genetic defect was an R40H mutation; R40 is found on the outer, convex side of the OTC trimer and its function is unknown. How this patient could live 56 years before undergoing an episode of hyperammonemia in spite of unmentioned but likely stresses like intercurrent illnesses, fasting episodes and trauma is difficult to understand. We speculate that such patients self-select a low protein diet because of subtle unpleasant symptoms and their OTC and other urea cycle enzymes are therefore down-regulated in amount by the low protein diet[(341,529)]. The low liver OTC activities (<10%) at onset of symptoms in some of these adult (>20y) late-onset patients (i.e. cases 17, 26, 36, 37, 52) may also be due to rapid enzyme degradation during the terminal illness with its release of cytokines.

The first patient recorded in the late onset male group still remains a puzzle. Levin's patient had 74% of normal activity at pH 8.0 using glycylglycine as the assay buffer and only 25% at pH 7.0 using Tris as the buffer. This enzyme showed *lower* K_ms for ornithine and for carbamyl phosphate versus the controls, i.e. greater affinity (Table 7-3). The effect of pH on activity of normal human OTC assayed in non-inhibitory triethanolamine-maleate buffer predicts that the activity at pH 7 would be 79% of that at pH 8([89]). The OTC activity in control livers at pH 7 in Tris was reported as 90% of that at pH 8 in glycylglycine by Levin[(487)]. Without gene studies in this family we cannot explain this puzzling mutation; no others like it have been reported (Tables 7-2, 7-3).

One remarkable and carefully studied case is late onset male patient 16, that of Hoogenraad[(499)]. The activity was 10% of normal at pH 7.4 in Tris buffer, the K_m for ornithine was normal and that for carbamyl phosphate was 15μM compared to 240μM for the control. The pH optimum was not measured. The cross-reacting material was 75% of that in the normal control liver. The size of the monomer on SDS-PAGE was 2000 daltons smaller than the normal monomer, suggesting that a truncated protein had been made. No gene studies have been done in this family, no other case like it with gene studies has been reported and no explanation is available to show how a deletion could increase carbamyl phosphate binding while impairing the overall catalytic rate.

Plasma Amino Acid Levels in OTC Deficiency

When the cases of ornithine transcarbamylase deficiency reported in the literature were reviewed from 1962 to 1999, those in which amino acid levels were reported and normal values given for reference are summarized in Table 7-7. The results are expressed as the percent of the *upper* limit of normal values for gln, ala, glu and lys because these are the amino acids which are likely to be elevated in OTC deficiency. Excess ammonia is first shunted to gln, ala and glu as temporary storage. For citrulline and arginine we expressed the results as % of the lower limits of normal because we expected them to be low. We also showed ornithine as % of lower normal because it was sometimes low. Ornithine does not accumulate in OTC deficiency because it escapes via ornithine keto-transaminase to glutamate or proline.

Among those *neonatal males,* gln was elevated 2-fold and citrulline was undetectable in 15/16 cases (Table 7-7). The assays were carried out within 1–4 days of onset in all but 3/17 of the gln values. Ala, glu and lys were 2 to 3-fold elevated. Arginine and ornithine results are not helpful, ranging from low to high. The increases in lysine are unexplained. Our review of the literature shows that 2-fold increases of plasma gln do not compare with the 15-fold increases in plasma ammonia in these neonatal cases but the two values were not always measured on the same plasma sample.

In the *late-onset male* cases plasma gln was only 1.9-fold increased while the scanty data on ala, glu and lys showed ~2-fold elevations. Citrulline levels were not as helpful as in the neonates, averaging 47% of lower normal, and ranging from zero in severely affected males to 129% of lower normal (Table 7-7). Many of these amino acid assays were done when acute symptoms had abated. Among the *symptomatic females* plasma gln assays were 2-fold elevated on average, ranging from 124 to 500% of upper normal (Table 7-7). Ala, glu and lys were often normal. Average citrulline levels were at the lower limits of normal but ranged from 0 to 385%. Arginine was only occasionally

Table 7-7. Plasma amino acid levels in neonatal-onset males, late-onset males and female patients with OTC deficiency

Amino acids	Gln	Ala	Glu	Lys	Arg	Citr	Orn
	% upper limit of normal				% lower limit of normal		
Early onset males							
mean	207	279	211	264	90	1.25	121
S.D.	110	297	27	239	73	5	72
range	80–398	58–958	180–232	140–690	20–186	0–20	59–245
n	17	9	3	5	6	16	5
Late onset males							
mean	186	229	216	234	137	47	126
S.D.	92	168		118	153	37	68
range	86–390	70–485	200, 232	70–390	0–540	0–129	20–198
n	17	6	2	6	10	14	6
Females							
mean	202	144	170	131	144	100	166
S.D.	85	82	104	80	109	113	127
range	124–500	38–461	35–290	46–344	0–500	0–385	40–540
n	39	24	11	15	30	31	24

low. Plasma amino acid levels in the 11 females with early onset were fragmentary and did not differ from those in the entire group. Citrulline values were 50% and 136% of the lower normal levels in the two cases with such data. Overall the amino acid patterns are not as helpful diagnostically as the elevations of plasma NH_3 or urinary orotic acid except when the plasma citrulline is undetectable, which predicts a severe OTC deficiency.

Plasma Amino Acid Levels from the Literature

Other investigators have reported on plasma amino acids from their own series of patients. Matsuda[530] surveyed 32 male patients in Japan and reported that in ten neonatal patients the average glutamine elevation was four-fold and the citrulline was essentially zero. In those patients with onsets from 29 days through 5 years the glutamate elevation was only 190% of normal and citrulline was 9% of the normal mean. In those five years of age and older glutamine was 320% and citrulline 43% of normal. Zhang et al.[531] reported a series of males between ten weeks and 23 years of age. In these patients the glutamine ranged from 140 to 540 percent of normal, the arginine values were all within the normal range, and the citrulline values were only low in four of the ten patients. Finkelstein[511] reported a series of males beyond the neonatal period. In eleven of these the average glutamine was 240 percent of the normal mean value. It appears that the glutamine level drawn during the acute stage or when the plasma ammonia is elevated is almost invariably abnormal[481]. The citrulline level, if it can be measured accurately, is frequently low and suggests OTC deficiency. The arginine is not helpful, usually residing on the low side of normal. Ornithine is neither elevated nor depressed on the average and is not of diagnostic usefulness. Occasional elevations of lysine are confusing and require exclusion of lysinuric protein intolerance. Methods must be used which prevent glutamine deamidation in plasma and a clear separation from glutamate. The correlations between plasma ammonia and glutamine were highly significant ($r = 0.77$) when measured in one patient over two years[481] but the correlation was poorer when plasma ammonia was normal. The ammonia was normal until the glutamine rose above 800μM (upper normal = 643) suggesting that plasma glutamine is more sensitive than plasma ammonia. Paired glutamine and ammonia analyses in 7 OTC patients over years of treatment showed a great variation in ammonia values as glutamine levels increased, with no linear relation between the two[729]. The authors explained that this may be due to saturation of glutamine synthetase followed by rapidly rising ammonia levels.

The other major reason why the measurement of amino acid levels is useful is in making a diagnosis of citrullinemia, argininosuccinic aciduria or argininemia when hyperammonemia is detected. Deficiencies of either carbamyl phosphate synthetase or OTC cause hyperammonemia and elevated glutamine levels without elevations of urea cycle amino acids like citrulline, argininosuccinate or arginine. In this situation measurement of orotic acid pays greater dividends because urinary orotic/orotidine levels are not elevated in CPS deficiency or in the rare deficiency of acetyl-glutamate synthetase, which results in secondary CPS deficiency.

Case Series of OTC Deficiency From the Literature

Several case series of patients with OTC deficiency have been published as literature reviews or personal series and they will be summarized here briefly. The first report was written by Bachman[532] in which he reviewed detailed case reports from the literature which included 22 females and 37 males, and combined these data with that from 15 females and 14 males which he had collected both in Lausanne and in collaboration with Columbo in Berne. He focused mainly on male variants because this category was less well known beyond the newborn age. In the neonatal period seizures occurred more often than impaired consciousness or coma. Hypo- and hyperthermia were noted occasionally. Respiratory distress and hyperpnea were exclusively seen in newborns before 15 days of age. Vomiting predominated in all age groups. Ataxia was an indicator of hyperammonemia mainly between ages two and five. During and after puberty behavioral symptoms such as confusion, disorientation, hallucinations and language problems were most common. He found that patients could not be accurately assigned based on the presence of alkalosis or absence of acidosis when differentiating them from organic acidurias. He proposed a 24-hour biochemical workup once hyperammonemia had been proved, including quantitative amino acid analysis in plasma and urine, orotic acid in urine, organic acids in urine and medium chain fatty acids in plasma. While these data were being collected he treated his patients with arginine, carnitine and glucose until a diagnosis was apparent.

Bachman also reported studies[532] to show how the sample size could affect the results of OTC activity assays in liver biopsies because of the mosaic pattern of OTC in the livers of heterozygotes. He found that 12 to 16 cells made up the largest diameter of patches of active or inactive OTC clones. He calculated that each mg of liver should sample between 370 and 880 clones of this size. Thus a 10 mg needle biopsy of liver should be a representative sample of the mosaic pattern of OTC positive and negative patches. However Glasgow[392] found that 5 mg pieces of liver from an affected heterozygote varied 10- to 40-fold in their OTC activities. Control liver samples varied 2.5-fold in OTC activities. Using our OTC assay[525] we found minor variations in activities from multiple samples of a rat or a normal human liver. Yorifugi[541] took 150–200 mg wet weight samples from a heterozygote liver and found four OTC activities ranging from 6.8 to 15 to 16.2 to 32.6 units/h per mg protein. Thus the poor correlations reported between severity of hyperammonemia or age of onset in *females* and liver OTC activities may be due in part to sampling errors in female mosaic livers, combined with wide variances in the reproducibility of the assays.

Matsuda[530] did a retrospective survey of all the male cases of OTC deficiency identified in Japan between 1978 and 1988. He found 10 patients with age of onset between 0 and 28 days (group 1), 13 with an onset between 29 days and five years (group 2) and nine with an onset of five years or more (group 3). The mortality in the neonatal group was six out of 10, in group two was two out of 13, and in group 3 it was 8 of 9. The mean percent of normal liver OTC activities (± SD) in these three groups were 3.7 ± 3.6%, 12.8 ± 7.0%, and 5.1 ± 2.9% respectively. The number of patients surviving with normal mental function was only one of four in the neonatal group, was nine out of 11 in group two and was normal in the only survivor of 9 in group three. During the first episode the mean values of plasma NH_3 in μM were 1175 in group one, 279 in group two and 605 in group three (upper normal = 100μM). Citrulline averaged 0 μM in group one, 3 μM in group two and 13.7 μM in group three. Urinary orotic acid in mg/gram creatinine averaged 427 in group one, 987 in group two and 895 in group 3 (the upper limit of normal being 5.2). The high mortality in patients with onset after age five is unexplained.

Zhang et al.[531] measured the OTC activity and Western blots of OTC protein in the livers of 11 patients with *neonatal* onset of OTC deficiency. He found that OTC activity varied from less than 0.05 to 0.74 μmol/min per gram of liver (the normal being 93.4 ± 6.3). The patients with less than 0.05 units all had severe hyperammonemia and coma. Those treated prospectively were stabilized and went on to liver transplantation and were alive and well by the last report. The patient with the highest activity (0.8%) in the liver had mild hyperammonemia and no coma. On Western blots of liver OTC, ten of the patients had no detectable monomeric bands whereas the patient with residual enzyme activity had a clearly visible band with a normal molecular weight of 36,000. The nature of the defects in these particular patients had not then been determined.

Over the years 1972–1988, Cathelineau, Briand and Rabier and their associates at the Hopital Necker-Enfants Malades in Paris performed extensive kinetic and protein studies of patients with OTC deficiency[383,449,452,471,508,514,529,533–536]. They systematically studied the effects of pH on K_ms for both OTC substrates[529,534] and on V_{max} and V_{max}/K_m looking for ionizations of functional groups at the active center or on the free enzyme[535] and identified a number of pH mutants with shifts in the pH optimum or pK of the ionizing group (normally histidine, pK ~6.6) at the active center[89] to higher pK values, always impairing ornithine binding. A review of 19 male cases in 1988 added the measurement of immunoassays (Western blots) and identified *seven* groups[536]. The first group of eight cases consisted of neonatal onset male patients who had no measurable activity at pH 8 or 9 and showed no cross-reactive material with either bovine or mouse anti-OTC antisera. The one group 2 patient had no activity but 1% of OTC protein. Two other children (group 3) had no detectable activity but 80% of control protein levels. Group 4 consisted of one child with 1% activity at pH 8 and 9, normal K_ms and 1% of protein. In group 5 three patients had 5% of normal activity at pH 8 and 9, normal Km's for ornithine and carbamyl phosphate, and 5% of control protein levels, resembling the spf^{ash} mouse (see Chapter 6). One pH mutant (group 6) had 0.5% activity at pH 8, but 30% at pH 9, a K_m for ornithine 5 fold above normal and 30% of control OTC protein. The last three children in group 7 had an activity of 20% at pH 8 and 60% at pH 9. Their K_ms for ornithine were 5–10 fold higher than normal and the amount of protein was 50–60% of controls.

We now know from studies of the genes in male patients how some of these defects can arise. Group-1 phenotypes can result from generation of a stop codon in pOTC at arg 92 (CGA to TGA)[(240)] or a splice defect and deletion of an exon[(273)]. Group-3 can result from an arg 94 conversion to threonine at the carbamyl phosphate binding site that gives almost no activity but nearly normal amounts of enzyme protein[(254)]. Group-5 is illustrated by a Spanish family with the same defect as occurs in the spf^{ash} mouse[(271)]; this results in a 5% level of functioning H117 protein due to a donor splice site error which adds 16 amino acids to exon 3, resulting in a monomer which is degraded. Group-7 phenotypes can result from a C-to-T mutation converting arg-277 to tryptophan[(242)]. Groups -2, -4 and -6 have no known genotypes that fit their phenotypes.

One of the best studied patients that shows how to separate OTC deficiency from Reye's syndrome in an adult male was that of Tallan[(500)]. A 21 year old male presented with the first onset of nausea, vomiting and coma. His brother had died in the neonatal period. The liver OTC activity was 8.4% of normal. Liver histology showed nonspecific reactive hepatitis. Electron microscopic studies showed a few breaks in mitochondrial membranes, crystalloid inclusions, bizarre shapes of some of the mitochondria and decreased dense bodies. Mitochondria were not massively swollen as occurs in Reye's syndrome. There was an increase in peroxisomes, not a decrease as is seen in Reye's. There were minimal fat particles. The orotic acid excretion was markedly elevated which is unusual in Reye's syndrome. One of Tallan's contributions was to analyze the free amino acids *in the liver* expressed as mmoles/100g of liver. There was a marked elevation of glutamine, glutamate and alanine; a normal ornithine level; absent arginine; elevated lysine and histidine; but citrulline was not reported. These results in the liver paralleled closely the results in the plasma in this patient. The overall findings were those of genetic OTC deficiency, not Reye's syndrome.

Saheki's patient[(537)] was remarkable because this male patient had a normally active mRNA by translation in a rabbit reticulocyte lysate system but no detectable immunoreactive protein in the liver. This suggested that the OTC-related protein was being synthesized from this mRNA but was being degraded too rapidly to be detected by any immunoreactive method. *In vitro* translation produced a pOTC protein with a normal molecular weight which could be transported into rat liver and kidney mitochondria and formed a trimer with a molecular weight that of mature OTC. Why the mature enzyme was degraded so rapidly *in vivo* was not clear.

Kodama[400,401] expanded on this theme by analyzing two female heterozygote patients who showed decreased amounts of OTC protein by immunoblotting. The mRNA from the first patient directed the synthesis of a small amount of OTC precursor of normal subunit size. In patient two the mRNA directed the synthesis of small amounts of two *in vitro* products, one of the usual 40,000 kDa and the other of 30,000 kDa. The normal sized pOTC of patient two was converted to mature-sized OTC by isolated rat liver mitochondria and must have been coded for by the normal X chromosome while the smaller product, coded for by the mutant X chromosome, was degraded during the incubation with the mitochondria.

Rowe[406] summarized the natural history and the differential diagnosis of OTC deficiency in 13 symptomatic female heterozygotes. Episodic extreme irritability occurred in 100% of these patients, episodic vomiting and lethargy in 100%, protein avoidance in 92%, ataxia in 77%, stage 2 coma in 46%, delayed physical growth in 38%, developmental delay in 38% and seizures in 23% (compare with our results in Table 7-4). 42% of the female members of these 13 families had OTC deficiency symptoms. Unfortunately, there was an average delay between onset of symptoms and diagnosis of 16 months. Five of these patients had IQ scores below 70 at the time of diagnosis. Rowe warned clinicians that they should be more aware of the symptoms of hyperammonemia, the episodic nature of these symptoms, the need to evaluate the family history and the urgency for an early diagnosis so that therapy can be initiated and thus prevent brain damage. Lysine was elevated in the plasma of some of these patients so that lysinuric protein intolerance had to be ruled out by a lysine-loading test if the OTC activity was greater than 50% of normal.

A review by Batshaw et al.[407] also made a number of the same points. They analyzed 17 kindreds containing 114 women who had a risk for heterozygosity. Sixty-one of these women were designated as heterozygotes by pedigree analysis, history of protein intolerance, protein tolerance tests or DNA probe studies. Eighteen percent of the 61 heterozygotes had experienced encephalopathic episodes and 9 of these patients died during these episodes. Batshaw made a plea that within such kindreds females at risk should be identified early and considered for long-term alternate pathway therapy (phenylbutyrate) and that they should be treated aggressively during hyperammonemic episodes to prevent death or brain damage. Drogari[504] made much the same point in his summary of six boys with OTC deficiency presenting beyond 28 days or in later childhood. Recognition of the subtle symptoms of hyperammonemia, particularly confusion, irritability or behavioral problems should lead a thoughtful physician to the measurement of plasma ammonia or urinary orotic acid. Drogari showed that with early recognition of hyperammonemia and phenylbutyrate therapy in these late-onset males the prognosis is relatively good.

Tuchman's group, formerly at the University of Minnesota and now at Children's National Medical Center in Washington,DC, has served for many years as a referral center for those wishing to confirm the diagnosis and mutations of OTC deficiency or to collaborate on studies of such patients. They have published reviews of their patients with OTC deficiency studied by the usual clinical and laboratory measures[538], by Western blots[531], by the biochemical and molecular spectrum[170], by comparing gene defects with clinical biochemical phenotypes[217], and by the clinical, biochemical and molecular spectrum of the disease[9]. In 2000 Tuchman collaborated with the groups at the University of Pennsylvania in describing the genotype spectrum of OTC deficiency[303]. A total of 317 families were investigated for possible OTC deficiency because of hyperammonemia and orotic aciduria in the absence of data suggesting other urea cycle defects, lysinuric protein intolerance or hyperammonemia, hyperornithinemia and homocitrullinemia (HHH) syndrome. Of these 317 families 157 (50%) had deleterious mutations found. Deleterious mutations were defined as those that do not function when transfected into cells which do not express OTC, that place a new amino acid at a site highly conserved or that is known to change catalysis or structure in OTC modeling. Liver enzyme analysis was diagnostic of OTC deficiency in 53 propositi but a deleterious mutation was only found in 41 (77%), a discrepancy thought to be due to mutations in noncoding regions. Among those males with neonatal onset, 31 missense and 6 splice site mutations were found. The phenotype of hyperammonemic coma within the first 5 days of life is stereotypic and is seen with 0-to-trace liver OTC activity, median peak plasma ammonia levels of 1420

μM and [^{15}N] ammonia incorporation into urea of 0.45–3.9% of normal. The mutations which cause the severe phenotype by their conclusions are as follows: large deletions or small out-of-frame deletions or insertions; termination mutations; missense substitutions in highly conserved domains within the active site or within the interior of the tertiary protein structure; or splice site mutations of the donor or acceptor consensus sequences. *Late onset* male mutations totaled 25; 23 were missense mutations and 2 were 3-base in-frame deletions. The phenotypes were variable, with onsets from 29 days to adult life or never (asymptomatic), with median peak ammonia levels of 400μM and labeled ammonia incorporations from 35 to 74%.

Mutations in *females* were also identified by SSCP screening and direct sequencing of the 10 exons and exon-intron borders plus cloning of PCR fragments into a plasmid so that the mutant and normal plasmids of heterozytes could both be sequenced. Twenty three missense and 5 splice site mutations were found. Four occurred at the same amino acids as neonatal males but none at the amino acids of late onset males. Twenty one amino acid mutation sites were unique. Relative frequencies of mutations in these 157 families revealed that substitutions account for 86%, splice sites for 13%, CpG dinucleotides for 31%, R141Q for 4%, R277W for 4% and all other mutations for 3% or less(539). Tuchman extended his report of 2000 by an extensive compilation of old and new mutations from his laboratory and those from the literature through 2002(661). He included 13 polymorphisms. Our Tables 5-1 and 5-2 include all of these mutations and polymorphisms with a few corrections.

Finkelstein(511) also looked at late onset ornithine transcarbamylase deficiency in their series of male patients. Late onset was defined as greater than 28 days of age. Probands were normal at birth but irritability, vomiting and lethargy later developed. The age of presentation ranged from two months to 44 years. As expected they showed high ammonia and glutamine, low citrulline, high urinary orotate excretion and decreased liver OTC activity ranging from 0 to 15% of normal. Patients older than 12 months of age usually presented with behavioral changes, decreased appetite, poor growth, mental retardation and protein avoidance. Precipitating factors included upper respiratory (viral) illnesses and dietary changes from human to cow's milk. Of the 21 male patients 12 survived and 10 are now being treated with alternate pathway therapy, i.e, phenyl-butyrate. They recommend that patients with a history that sounds like hyperammonemia do not undergo the one-gram/kg protein tolerance test but be tested with allopurinol that avoids an acute crisis.

Symptoms of OTC Deficiency Precipitated by Valproic Acid

Six female heterozygotes and two late-onset males developed symptoms and hyperammonemia after institution of valproic acid (sodium valproate) therapy for a seizure disorder. The female cases in Table 7-1 are numbered 41, 52, 62, 69, 87 and 107. the male cases in Table 7-3 are numbered 12 and 34. After starting therapeutic doses of valproate the onset of symptoms and hyperammonemia began in one day in two cases, and in 2, 6 and 7 days in the 5 females where the history was adequate to judge. The onset in male case 34 (Table 7-3) was 5 days after valproate was begun and after 2-6 weeks in case 12. Symptoms consistent with OTC deficiency occurred *before* the valproate therapy in female 52 (one year duration), in 62 (one week), in 69 (since birth) and in 107 (for 11 years). The two males had no warning history. Two of the 6 females died within 5 days in spite of stopping the valproate. Both early onset males died and their post-mortem liver OTC activities were only 2.8 and 6.2% of control values. The two female post-mortem liver activities were 6.2 and 15% of controls; only one survivor had liver activity measured on a biopsy and it was 32% of controls.

The reasons why valproate impairs ammonia removal by the liver are unclear. Valproic acid is 2-propylpentanoic acid. Like other odd-chain fatty acids, it is converted in the outer mitochondrial membrane to valproyl-CoA, carried through the inner membrane as valproyl-carnitine, reformed to valproyl-CoA and then by β-oxidation converted to two molecules of propionyl-CoA, thence to methylmalonyl-CoA, succinyl-CoA, succinate and by the usual Krebs cycle steps to acetyl-CoA. Much less ω-oxidation also occurs.

When rats were given valproate 100 mg/kg intraperitoneally and mitochondria isolated 2 hours later citrulline formation was only 12% of controls using succinate as an energy source; no inhibition occurred when glutamate was the source(547). The activities *in vitro* of N-acetylglutamate synthetase (NAGS), CPS-1 and OTC were not affected but N-acetylglutamate (NAG), the essential co-factor of CPS-1, was reduced by 25%. Permeabilized mitochondria allowed free ATP uptake but citrulline formation was still impaired. Coude et al.[547] postulated that valproyl-CoA was competing with acetyl-CoA for formation of NAG. Subsequent studies in isolated rat liver hepatocytes incubated with alanine and ornithine[548] and valproate 0.5–5 mM confirmed that NAG was reduced to 18% of controls by 2 mM valproate. Also reduced by 50% or more were acetyl-CoA, aspartate, citrulline, glutamate and urea synthesis. Formation of valproyl-CoA was thought to be the cause of the low acetyl-CoA and NAG in these cells. When intact mice were given 13 mmol/kg of alanine and 4 mmol/kg of ornithine, marked increases in liver NAG, glu, gln, asp, citrulline, argininocuccinate, arginine and urea occurred[549]. Pretreatment with valproate 200 mg/kg lowered the expected increases in NAG, glu, gln and asp and lowered acetyl-CoA but not citrulline, argininosuccinate, arginine or urea. Blood ammonia after alanine and ornithine alone did not increase but when valproate was given it equaled 168% of controls. The decrease in NAG was not sufficient to impair activation of CPS-1 given the K_a value of 11 μM for NAG in phosphate buffer[550]. The theory that CPS-1 function was controlled by the mitochondrial level of NAG had been questioned before[551].

Methylmalonyl-CoA inhibited purified CPS-1 (K_i = 5mM)[552] and propionyl-CoA inhibited CPS-1 in human liver homogenates[553] by lowering V_{max} but not affecting the Michaelis constants for ammonium, bicarbonate or ATP. However, valproate given for 9 days at 0.05, 0.15 and 0.25% in drinking water to *spf* male mice with 11% of normal OTC activity caused *no* inhibition of CPS-1 when assayed with optimal NAG in liver homogenates but CPS-1 was surprisingly induced[554] almost two-fold. Urinary orotate excretion did not change indicating no effect of the doubled CPS-1 content on production of carbamyl phosphate within the mitochondria. Yet valproate caused 33% mortality in the *spf* mice but none in the control mice. The liver histology in controls given valproate showed occasional single cell necrosis, infiltration with neutrophils and slight fatty vacuolization. *Spf* mice showed necrosis in all livers, centro-lobular necrosis in 20%, more neutrophils, more fatty change and enlarged mitochondria. *Spf* mice are the best model for humans with OTC deficiency given valproate. It appears that valproyl-CoA and its products propionyl-CoA and methylmalonyl-CoA compete for CoASH and reduce NAG synthesis and furthermore inhibit CPS-1 by binding to the NAG binding site. High doses seem to cause direct liver toxicity.

In adult seizure patients taking standard doses of valproate 30–50% developed 5–300% increases in plasma ammonia but with few symptoms[555,556]. Children were more likely to develop higher ammonia levels and symptoms especially if less than age 2, if they are taking other anti-seizure drugs or develop valproate levels above 100 μg/ml[557,558].These elevations in plasma ammonia can be prevented or reversed by treating with oral L-carnitine at 50–100 mg/kg per day[559,560]. Valproate can cause a secondary deficiency of carnitine, as can chronic hyperammonemia in OTC deficiency. Acyl-CoA esters formed in excess in mitochondria are excreted as acylcarnitine esters in the urine, depleting body stores of free carnitine. Patients consuming a low meat, vegetarian diet such as those with symptomatic OTC deficiency are susceptible to carnitine deficiency[559].

Valproate elevates serum glutamic-oxaloacetic and -pyruvic transaminases in up to 44% of patients on usual doses[561,562] but these elevations do not reliably predict which patients will develop hyperammonemia or rarely develop acute liver failure due to submassive necrosis or microvesicular lipid deposits without much necrosis[562,563]. The cause of massive necrosis could be by an immune mechanism or by direct toxicity of valproate as seen in spf mice. Patients with CPS-1 deficiency[564] or multiple acyl-CoA dehydrogenase defects[565] can also have symptoms and hyperammonemia precipitated by valproate treatment. Four of five patients with valproate-related hyperammonemic encephalopathy (VHE) had elevated glutamine levels in serum or cerebrospinal fluid[730]. VHE is more likely when the patient is also given topiramate, another drug for partial seizures[731]. O'Neill et al.[732] found 30 cases of VHE since 1979. Our review of case studies of this entity indicates that rarely is a urea cycle deficiency or an organic aciduria screened for when ammonia levels rise to symptomatic levels while serum valproate assays are in the therapeutic range. The main lesson to be learned from this valproate story is to seek in the history any clues suggesting OTC deficiency or any urea cycle enzyme deficiency or organic aciduria before beginning valproate treatment and to watch for symptoms of hyperammonemia in the first 2 weeks of therapy. Before starting valproate therapy it may be wise to check a urine organic acid profile and to check blood carnitine levels periodically.

Rett Syndrome and Hyperammonemia

In 1966 Andreas Rett reported a dementing syndrome in 32 girls with onset between 6 months and 2 years[566] and described elevated serum ammonia levels in all 17 of the girls in whom it was measured. The ammonia method used serum and an amino acid analyser that are known to give falsely elevated results for ammonia. Hagberg in 1983 reported his own series of 16 Swedish girls with Rett syndrome and added 19 more cases from French, Portuguese and Swedish sources[567]. Only 4 of 23 had elevated ammonia levels, two between 55–99μM and two at 650 and 670μM. The ammonia methods were not described. Plasma amino acid levels were not abnormal in either series[566,567]. The diagnosis of Rett syndrome was based on a set of 8 clinical criteria[568] as follows: female sex; essentially normal development for 6–18 months; normal head circumference at birth with lack of head growth from 6m to 4y; early loss of achieved abilities and signs of dementia; loss of hand skills over 1–4y; hand-wringing-clapping stereotypic behavior over 1–4y; gait and truncal apraxia/ataxia over 1–4y; and diagnosis tentative until 3–5y.

A protein load (2 g/kg) was given to 3 girls with the Rett syndrome who had fasting plasma ammonia values of 67,67 and 126μM (normal<50); one girl developed an ammonia of 138μM (<90) 2h after the load[569]. Because no elevations of urinary orotate occurred the authors suggested that a deficiency of CPS, not OTC was present. A study of 12 girls after alanine loading revealed elevated urinary orotate excretion in two and both of their mothers excreted excessive orotate after an alanine load[570]. Another study with alanine loads produced elevated orotate excretions in 2 of 5 girls whose findings met Rett syndrome criteria. Orotate excretion was abnormal in the mother of one girl[571]. Another study using alanine loading found elevated urinary orotate excretion in one of 8 girls with apparent Rett syndrome[572]. A report of alanine loading of 10 girls who fit the standard criteria for Rett syndrome[573] revealed two with elevated fasting ammonias and a 2-fold increase after alanine. One of the two girls was receiving valproate treatment. A third girl had an elevated urinary orotate after alanine loading.

Hyman and Batshaw had reported an 8y old girl with OTC deficiency who had severe progressive mental retardation, ataxia, seizures, hand-wringing, and a non-growing head circumference[574].

They described the patient as a *phenocopy* of Rett syndrome and postulated that those apparent Rett cases in the literature with hyperammonemia and/or elevated urinary orotate excretion actually represented OTC deficiency, mimicking Rett syndrome. The problem was solved when the gene for Rett's was localized at Xq28. Mutations at this site in a methyl CpG binding protein (MeCP2) were associated with Rett syndrome(575). Male cases rarely occur but usually die in utero. In Rett cases defined by this genotype no hyperammonemia or OTC deficiency has been reported(576).

Commentary. Between 1962 and 2002 we have located 118 reported cases of OTC deficiency in female heterozygotes, 67 cases of male neonatal onset and 82 cases of male late-onset (after 28 days). Summaries of the clinical and laboratory findings in these 267 cases are listed in Tables 7-1 through 7-5. In all three categories a high index of suspicion for hyperammonemia is needed in order to do a prompt diagnostic evaluation and begin effective therapy to prevent death or severe brain damage in OTC deficiency as well as in other causes of ammonia intoxication. The efficiency of diagnostic tests has been defined. The clinical syndromes and laboratory findings for neonatal and late-onset males and in heterozygous females we describe are confirmed by case series by other investigators. Correlations between phenotypes and genotypes are illustrated when possible. The role of valproic acid in precipitating hyperammonemia in OTC deficiency is discussed. Confusion between Rett syndrome and OTC deficiency has been resolved.

8

Diagnosis and Treatment of OTC Deficiency

We have dealt with a number of issues in the diagnosis and treatment of OTC deficiency in previous chapters. Chapter 5 reviews the use of RFLPs[226,228,233] and DNA[225,227,232–234,240] analyses in diagnosis and introduces the use of allopurinol in testing for the carrier state[241]. Chapter 7 discusses carrier detection with protein[393], alanine and allopurinol loading tests and some aspects of treatment of hyperammonemia by hemodialysis[376], diet[394], carnitine[480,543], alternative pathways of nitrogen excretion[492] and by liver transplantation[414,4333,437,438,443]. The role of imaging methods is noted for computed tomography (CT) and magnetic resonance (MR) scans[403,422,425,435]. Measurement of orotic acid/orotidine in urine is discussed[523,524] and also methods for assay of OTC in tissue samples[525,526,539,540]. In this chapter we expand on all aspects of diagnosis and treatment.

Diagnosis

The diagnosis of neonatal hyperammonemia should be considered in any newborn with failure to feed, somnolence, lethargy or coma, and especially respiratory alkalosis. Sepsis is usually considered the cause of such a syndrome. Therefore blood or plasma ammonia assays should be obtained as part of any sepsis workup[577]. Late-onset hyperammonemia presents with loss of appetite, protein avoidance, cyclical vomiting, lethargy, ataxia and behavioral abnormalities[578] such as hyperactivity, night restlessness, biting, self-injury, delusions, hallucinations and psychosis. Adults manifest mostly psychological and neurological symptoms and signs.

The keystone of diagnosis is the measurement of plasma or blood ammonia in a patient with symptoms and signs consistent with hyperammonemia. The best methods review in our opinion is that of Huizenga et al. in 1994[579]. It compares the range of normal values in venous plasma for the glutamate dehydrogenase method for ammonia assay, three examples of ranges being 11–55, 11–60, 21–99μM; the automated Ektachem diffusion method (16–35μM); and a fluorimetric method (17–58μM). The Blood Ammonia Checker (BAC II) reported a normal range of 28–34μM[580]. The care necessary in collecting and rapid cooling of blood, separation of plasma and ammonia assay within 90 minutes was stressed. The review compares many old and new methods and illustrates the need for each laboratory to establish its own upper limit of normal for the method it employs. A more recent review of ammonia methods is that of Barsotti[581] but no upper limits are given for the various methods. If the newborn is premature and has respiratory distress *before* 24h of age, the diagnosis is more likely to be transient hyperammonemia of the newborn (Figure 8-1)[582].

The next step in evaluation of hyperammonemia is to obtain qualitative measurements of *urinary organic acids* (Fig. 8-1)[577], usually by gas chromatography-mass spectrometry[583,584](GC-MS). If abnormal amounts or kinds of organic acids are found this leads to consideration of congenital lactic acidoses, organic acidemias or fatty acid oxidation defects. If urinary organic acids are normal, the next critical measurement is *plasma citrulline* along with other amino acids by an amino acid analyser[585] or by blood spot amino acid analysis by tandem mass spectrometry[586]. Urinary amino acids should be measured to confirm lysinuric protein intolerance (LPI) or urinary homocitrulline as part of hyperornithinemia-hyperammonemia-homocitrullinemia (HHH) syndrome (Fig. 8-1). If plasma citrulline is absent or trace in amounts the next critical measurememt is *urinary orotate* and/or orotidine, which was first reported in OTC deficiency as elevated by Levin[371]. Urinary orotate and orotidine are elevated whenever carbamyl phosphate, made in mitochondria by CPS-1, is not rapidly utilized by OTC for citrulline formation, escapes from mi-

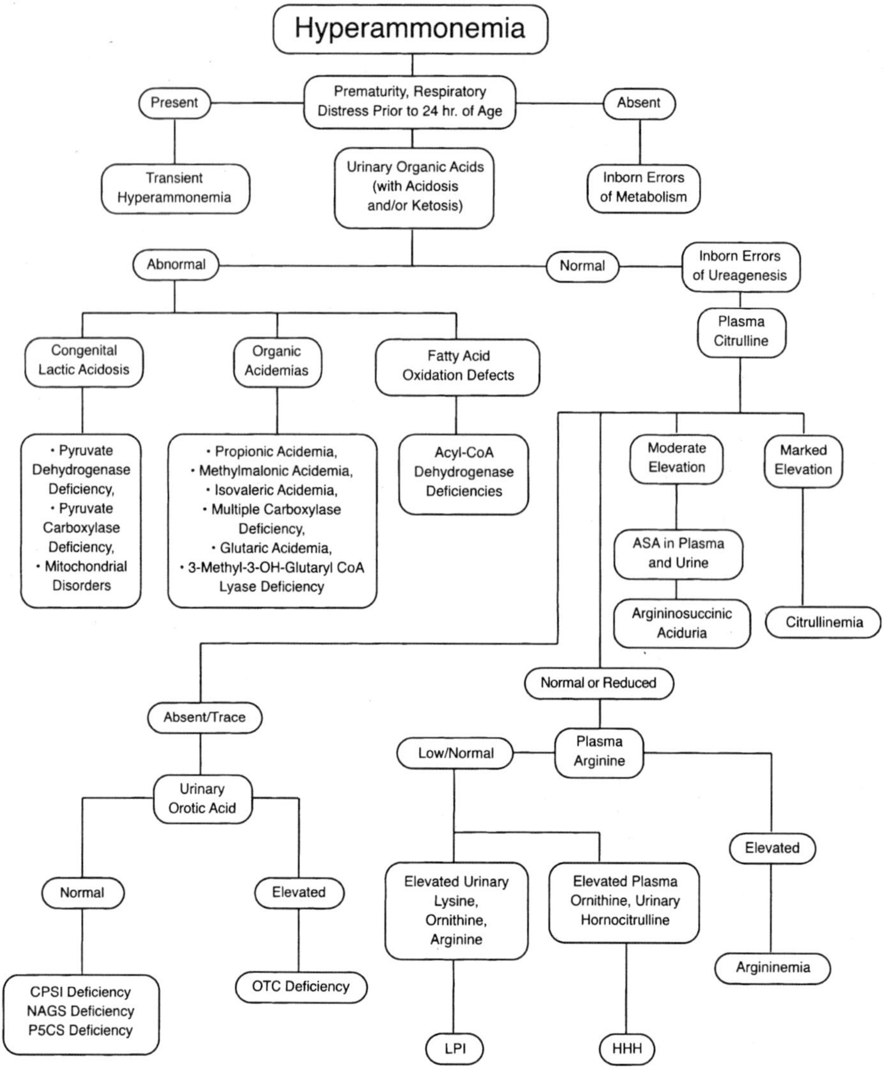

Figure 8-1. Flow chart for the differential diagnosis of congenital hyperammonemia[578]. Abbreviations: LPI, lysinuric protein intolerance; HHH, hyperornithinemia-hyperammonemia-homocitrullinemia syndrome; P5CS, pyrroline-5-carboxylate synthase; NAGS, N-acetylglutamate synthetase; CPS1,carbamyl phosphate synthetase. (Copyright 2001. Reprinted with permission from Elsevier.)

tochondria into the cytoplasm where it accelerates carbamyl aspartate synthesis and subsequently orotate and orotidine formation. Besides OTC deficiency orotate in urine is elevated in argininosuccinate synthetase (AS) deficiency, occasionally in argininosuccinate lyase (AL) deficiency, in arginase deficiency, HHH syndrome, lysinuric protein intolerance (LPI)(586), and in mitochondrial diseases(587), uridine monophosphate synthetase deficiency (classic orotic aciduria)(588), benign persistent orotic aciduria(588) and in normal pregnancy(589). These other causes of orotic aciduria can be separated from OTC deficiency by the clinical picture and automated plasma amino acid analysis or filter paper blood spot analysis of amino acids by tandem mass spectroscopy(586). Another cause of hyperammonemia and orotic aciduria is Δ^1–pyrroline–5-carboxylate synthase (P5CS) deficiency (Fig. 8-1)(590). This enzyme catalyzes the reduction of glutamate to P5C, one isoform being active in the gut mitochondria. Ornithine aminotransferase produces ornithine from P5C that couples with CPS-1 and OTC to produce citrulline which is converted to arginine, mainly in the kidney and gut. The other isoform produces proline from P5CS in many tissues. This syndrome of neurodegeneration, joint laxity, skin hyperelasticity and subcapsular cataracts is associated with hyperammonemia, hypoornithinemia, hypocitrullinemia, hypoargininemia and hypoprolinemia. The ornithine deficiency in the mitochondria causes the fasting hyperammonemia and orotic aciduria which improves with a protein meal containing arginine(590).

The final proof of OTC deficiency is an assay of OTC in liver or intestinal mucosal tissue. However, OTC assays in tissues are fraught with possible errors. Surgical or autopsy wedge biopsies sample enough clones of normal and OTC deficient cells to avoid sampling errors in female heterozygotes. Needle biopsies must contain 20–40 mg wet weight of liver to avoid sampling errors(392,525,532). A Klatskin or Tru-cut needle contains this liver weight in a full tray but a size-16 suction biopsy needle (Menghini-type) often requires two samples to total 20–40 mg. The baseline of activity per mg protein in needle biopsies is likely to cause low results because of inclusion of hemoglobin and blood proteins, even after blotting of blood(525). An optimal baseline in needle biopsies is DNA(525) but it is rarely used. The many other factors affecting enzyme activities in human liver and in liver biopsies are listed in a review by the author(591).

Carbamyl phosphate and ornithine also form citrulline by a *non-*

enzymatic reaction which can amount to 5–10% of citrulline generated by low activities of OTC[89,141]. A blank reaction, which corrects for this problem and for endogenous citrulline and urea in the usual colorimetric reaction for citrulline[89,525], employs a boiled homogenate sample subsequently incubated with the substrates. Boyde[592] and our laboratory[525] found that a roughly equal mixture of concentrated H_2SO_4 (250ml/l) and concentrated H_3PO_4 (200–250ml/l) resulted in higher absorption values in the reaction with diacetyl monoxime and ferric ion plus antipyrine for 0.1 μmole of citrulline (E_{464}= 1.1) or with thiosemicarbazide (E_{530}=1.04) than the absorption with H_2SO_4 plus 5% acetic acid[592]. To insure linearity and proportionality in an OTC assay at least 2–3 dilutions at one incubation time or 2–3 points of time are needed[525,540].

These conditions can be applied to the assay of OTC in *human serum* as a test for liver injury, but preincubation with highly active, pure urease is necessary to remove urea interference[141]. There is no correlation between serum and liver activities of OTC. The same OTC assay methods have been used with small intestinal (or rectal) mucosal biopsies[593,594] but the intrinsic activity in duodenum is approximately 24% of that in liver. The diagnosis of OTC deficiency in the liver is reflected in the small intestine and rectal mucosa. OTC activity is found in very low levels in white blood cells, granulocytes or mononuclear cells[595,597] but there is no correlation with liver OTC values[596].

Other methods reported for OTC assays in liver include incubation with ^{14}C-ornithine and isolation of ^{14}C-citrulline on thin-layer chromatography[598], the spots of which were cut out and counted. Another method uses a standard assay with citrulline being separated on a chromatosheet covered with ion exchange resin and read as ninhydrin spots by videodensitometry[599]. An enzyme–linked immunosorbent assay (ELISA) was developed using an antibody against purified bovine OTC[600]; it was more sensitive than a colorimetric assay in serum. Another method used aliquots from a standard assay, which were boiled and the supernatants applied to a capillary electrophoresis apparatus where the citrulline peaks were read at 200 nm[601].

To increase sensitivity so that kinetics might be measured in tissues with very low OTC activity, radiochemical methods were developed, first using the conversion of ^{14}C-carbamyl phosphate to ^{14}C-citrulline[450,597]. The methods are more complex than the colorimetric ones described above, requiring column isolation of the ^{14}C-citrulline. The better radiometric methods employ L-[1-^{14}C]ornithine[526,598] but the separation of ^{14}C-citrulline was done by HPLC, the "cold" citrulline detected by absorption at 205 nm and the radioactive peaks by a radioactivity flow monitor[526]. The latter method was able to measure activity as low as 0.1μmol/min per g liver but 5 mg of liver tissue was required to do so.

When tissue OTC assays are within the broad range of normal in asymptomatic heterozygotes and loading tests are negative, measurement of rates of urea synthesis can be done using isotopes, measuring ^{14}C-acetate conversion to urea or ^{15}N-ammonia conversion to urea and glutamine[601,602]. ^{15}N-glutamine formation increased in asymptomatic heterozygotes and the ^{15}N-urea/^{15}N-glutamine ratio separated them from normal females[604] and correlated well with clinical severity in affected patients. Heterozygotes with normal liver OTC activities can be identified by analysis of all 10 exons and exon-intron borders by SSCP after amplification of genomic DNA by polymerase chain reaction[243] with a sensitivity of 80% for detecting a mutation in the affected female allele[303].

The most widely used colorimetric method for measurement of both orotate and orotidine is that of Harris and Oberholzer[523], based on bromination of the compounds followed by color development with p-dimethylaminobenzaldehyde. With pre-purification of urine on a strong cation exchanger column and appropriate blanks, this method gave a normal range in adults of 0.4–1.2 μmol/mmol of creatinine. Reference values for different ages showed a decline in total orotate + orotidine excretion from 2 weeks to 1 year (mean 2.37 mmol/mol creatinine, to 1–10y (1.82) to >10y (0.74). The methods which have superceded this are based on isocratic HPLC with a strong anion-exchange HPLC column that separates orotate from orotidine[605]. This system also measured orotate in serum[606]. Good results were also obtained for orotate by pre-purification of urine on a strong cation exchanger column and application to an HPLC column with tetrabutylammonium as a counter-ion[607]. Sebesta[608] found that the best pre-purification method was Dowex-1 column chromatography and elution of orotate and orotidine with HCl. The isocratic HPLC system (65% formic acid plus 35% methanol, pH 2.3) well separated orotate and orotidine from interfering compounds. Rimoldi[609] developed a stable isotope dilution assay using the same urine processed for organic acid analysis. To a urine sample was added 50 nmol of 1,3[$^{15}N_2$] orotic acid. The mixture was then acidified and saturated with NaCl, extracted with tetrahydrofuran, and twice extracted with ethylacetate and diethyl ether. Trimethylsilyl derivatives were analyzed by GC-MS. A similar method can be applied to dried filter paper urine samples and normalized to urinary creatinine concentrations(610).

Fasting or random urinary orotic acid/mg creatinine ratios do not always detect symptomatic heterozygotes unless they are in an acute hyperammonemic episode. Asymptomatic heterozygotes may have normal urinary orotate excretion even when given a protein-loading test. Studies in *spf* mouse heterozygotes where protein intake was controlled on a 22% protein diet showed that urinary orotate/mg creatinine ratios were elevated only when liver OTC activities were less than 50% of normal(611). We confirmed this problem in spf^{ash} heterozygotes where 5 of 16 had basal ratios within the range of 8 normal female mice(612).

The stable isotope GC-MS methods are said to be more sensitive and specific than HPLC methods(610), but those who find orotidine excretion to be better than orotate as a discriminant in loading tests use HPLC alone(241). Actually, the poor reproducibility of the Jaffé reaction for creatinine, and the change with age of normal creatinine excretion may account for more variance of the ratio of orotate or orotidine/mg creatinine on random or loading tests than the actual orotate or orotidine excretions, leading to use of 24h urine outputs without creatinine being involved(613). A simple test for urinary uracil, which like orotate is excreted excessively in OTC deficiency, was able to separate normal females from heterozygotes in spot urines without allopurinol loading, but the series was small and the findings need to be confirmed(733).

Loading Tests

In the 1960 and 1970s NH_4Cl loading tests were done on possible female carriers or asymptomatic siblings measuring plasma ammonia (Table 7-6) but were associated in some subjects with nausea, vomiting and high ammonia levels. Later protein or alanine loading tests were employed(393,614,615) measuring ammonia or orotic acid excretion (Table 7-6). False negative protein loading tests in heterozygotes[616] and false positive results in oral alanine loading tests[612] were reported. These loading tests also could cause hyperammonemic episodes[241]. A revised protein load (fat-free chicken breast) giving 35g protein/m^2 over 30 min and collecting urine for three 2h periods was reported by Potter et al.[618], expressing the results as the ratio of μmol orotic acid/mmol creatinine in the 2–4h/0–2h periods. Skim milk was found to contain orotic acid and to give a false positive test. Creatinine analysis using the Jaffé reagent was falsely elevated by acetoacetate (found in ketoacidosis) but not when alkaline picrate was used[613]. The Harris colorimetric method for urinary orotate+ orotidine gave the same results as the stable-isotope internal standard method of Rimoldi[618]. This improved protein load test was positive in 16 obligate heterozygotes and negative in 18 control subjects and 6 possible heterozygotes shown later not to carry the family DNA mutation. Side effects included one subject with headache. Four heterozygotes were unable to finish the meal but still showed positive results. The test was not useful in children who could not ingest the large protein load rapidly.

The *allopurinol* test for OTC deficiency was described by Brusilow and colleagues[241]. It relies on the conversion of allopurinol to oxypurinol ribonucleotide which inhibits orotidine monophosphate decarboxylase, preventing conversion of orotidine 5′-phosphate to uridine 5′-phosphate in the pyrimidine synthetic pathway. Orotidine and orotate accumulate and are excreted in the urine. If there is a block in disposition of mitochondrial CP, it will escape into the cytoplasm and stimulate carbamyl aspartate synthesis, a precursor of orotate and orotidine, and results in increased excretion of these compounds. Allopurinol amplifies this excretion. A 300 mg dose of allopurinol in 24 obligate heterozygotes increased urinary orotidine excretion in 23 during at least one of four 6h collection periods compared with the small increases in 25 normal women[241]. The sensitivity was 96% and specificity 100%. Urinary orotate by the HPLC method used was a poorer discriminant. Sebesta[608] found that his modification of an HPLC system gave orotic acid excretions greater than orotidine in controls, and their results in 23 female volunteers best fit a log normal distribution during three 8h collections. A cutoff above 2 S.D. was recommended for both orotate and orotidine. Diet was not controlled for protein or pyrimidine intakes. Burlina[619] established dose levels of allopurinol and normal values for children from 6m to 17y of age, using a pre-purified urine and a reversed phase HPLC. Only orotic acid was measured in four 6h collections because orotidine eluted with interfering compounds. Burlina has published doses of allopurinol adjusted for body weight that work well from infants(<1y) to adults[(743)].

Arranz[(620)] used a pre-purified urine sample and anion-exchange HPLC, measuring both orotate and orotidine. Orotidine showed better performance on Receiver Operator Characteristics (ROC) curves than orotate. The level of protein intake during the four 6h collections affected test results and as a result their female controls were asked to consume at least 2.5 g/kg of protein during the test. Log normal distributions of control values were found. At a cut-off of 8mmol orotidine/mol creatinine in children with mild OTC deficiency, the peak values gave a sensitivity of 1.0, a specificity of 0.92 and accuracy of 0.95. As expected, occasional heterozygote females with mild OTC deficiency showed negative results with both protein loading and allopurinol tests[(518,620)].

Neuroimaging, EEG and Pathology in Liver and Brain

Computerized tomography during *acute* hyperammonemia in neonates or late onset children or adults showed brain swelling with small ventricles and low density of white matter[(402,448)], enhancement of pial vessels after contrast injection and low density over one hemisphere[(435)]. CT scanning, however, may show a normal scan if done before brain edema develops. Hemiparesis was associated with low density of white matter in the left frontal lobe[(402)]. Longer survival in children or in adults after severe episodes resulted in scans showing cerebral atrophy, variable in location, in some cases involving the cerebellum but sparing the pons and basal ganglia[(402,621)]. Stroke-like episodes were associated with CT changes of infarction of grey and white matter[(424)]. Magnetic resonance imaging (MRI) also showed brain edema early, enhanced leptomeningeal vessels with gadolinium contrast and basal ganglia involvement[(435)]. An affected heterozygote showed MRI changes of increased T2 signal in the pons and midbrain, consistent with central pontine and extrapontine myelinolysis[(432)]. During acute hyperammonemia in two females proton magnetic resonance spectroscopy (MRS) of the head showed marked accumulations of glutamine in the brain[(425)], up to four times normal levels; the glutamine is thought to play a role in the brain edema, especially of astrocytes where glutamine is synthesized.

Electroencephalogram (EEG) findings in newborns with hyperammonemic coma showed multifocal independent spike-and-sharp-wave discharges, often with clinical seizures(485,622). Later when ammonia levels were normal moderate amplitude theta-delta activity predominated.

Liver pathology in OTC deficiency was essentially normal on light microscopy(519,623). By electron microscopy the findings were normal mitochondria, a slight increase in number of microbodies plus clusters of "pale cells" found by others to be due to free cytoplasmic glycogen, not microvesicular fat droplets as were present in most other hepatocytes(624). Nuclei often contained glycogen. A male newborn with the G299E mutation and 2% OTC activity showed hepatocytes with minimal lipid and much glycogen. Some mitochondria were pale on E.M. with short cristae, with a volume increased to 141% of normal. Immuno-gold labeling of OTC showed the enzyme increased in cytosol and in mitochondria. Degenerating mitochondrial-like dense bodies were numerous, but lacked OTC or CPS-1 and were not peroxisomes(625).

Neuropathology in the brain of a male infant who died in coma at 3 days showed softening of periventricular white matter, ischemic neurons in this area and in the striatum and thalamus. Alzheimer type II astrocytes were seen in affected areas(626). A male infant (not included in Chapter 7) had the first symptoms of hyperammonemia at 13h of age, a peak plasma ammonia of 2200μM, a slow response to therapy and died on the 17th day(627). The brain at autopsy showed "multiple, bilateral, symmetrically placed cystic lesions at the cortical gray-white junction." These lesions were infarcts with loss of neurons and formation of gliosis, possibly due to high intracranial pressure and poor cerebral perfusion. A two-year old girl who was severely retarded since age one showed a small brain with flattened gyri and shallow sulci over the cortex, and a shrunken cerebellum and thalamus. A massive neuronal loss and gliosis and Alzheimer type II astrocytes were seen in affected areas(626). One year after severe neonatal onset of OTC deficiency a male infant died and the brain showed shrinkage, cystic changes, astrogliosis of white matter and degenerate axon terminals from Purkinje cells in the dentate nucleus of the cerebellum, so-called grumose degeneration[728]. Overall the results of severe, prolonged hyperammonemia on the young brain are devastating.

Treatment

The author has not personally cared for patients with OTC deficiency. Therefore in this section we will rely on the experience of those who have. We will refer in particular to the Proceedings of a Consensus Conference for the management of patients with urea cycle disorders, published in 2001 as a supplement to J. Pediat. 138:S1–S80. Summar and Tuchman[578] and Summar[628] reviewed strategies for the management of neonatal urea cycle disorders. After recognition of hyperammonemia, early supportive care is critical: intravenous access, careful hydration with 10% glucose in quarter normal saline, endotracheal intubation and transfer to a neonatal intensive care unit for care by a team of specialists[578,628]. Some form of dialysis is needed urgently, the most effective according to Summar being hemodialysis driven by an extracorporeal membrane oxygenation pump[628] which achieves ammonia clearances of 170–200 ml/min. Continuous venovenous hemodialysis (CVVHD) gave the highest ammonia clearance and stable hemodynamics compared with continuous arteriovenous hemodialysis or standard hemodialysis[734]. Outcome was determined by the coma duration *before* dialysis. Continuous venovenous hemodiafiltration at a countercurrent dialysate flow of 5 ml/min achieved a clearance of 19.4 ml/min per 1.73 m^2[629] and has been modified in technique to give safe and effective treatment of hyperammonemia[735] but is not as efficient as that recommended by Summar[628]. Standard hemodialysis is less effective and homeostasis is difficult to maintain if continued for more than a few hours[628], especially in small infants. Peritoneal dialysis gives such low ammonia clearances that it is not usually recommended.

Alternative pathways for nitrogen removal have become mainstays of acute and long-term treatment[630,631]. One pathway consists of sodium benzoate (SB), which is activated by CoA to benzoylCoA that reacts with glycine (and alanine) to produce benzoylglycine (hippuric acid)and benzoylalanine respectively, which are rapidly excreted by the kidneys. Glycine is replaced from glycine stores and from glutamate via serine[632] in animals but studies in humans are lacking. Excessive doses of benzoate can be toxic. A more effective pathway occurs in human liver and kidney and is carried out by an acylating enzyme that activates phenylacetate (PA) to phenylacetylCoA. Then a second enzyme joins PACoA to glutamine to form phenylacetylglutamine (PAG)[633], which is rapidly excreted in the urine, thus removing 2 moles of N for every mole of PA. Loading doses of SB and sodium phenylacetate (SPA) intravenously are 250 mg/kg of each over 90 min and a maintenance infusion of 250 mg/kg of each over the next 24h. Arginine becomes an essential amino acid in OTC deficiency[635]. Supplying arginine helps replace mitochondrial ornithine to optimize remaining OTC function. Its loading dose is 600 mg/kg[628], and this dose is repeated every 24h. Doses of sodium benzoate, sodium phenylacetate and arginine for young and older children are given in Table IV of reference 628. Citrulline by nasogastric tube at 170 mg/kg per day replaces intravenous arginine. Phenylbutyrate, used enterally only, is converted to PA and avoids the unpleasant smell and taste of PA[634].

Long term management of diet, medications, clinical and laboratory monitoring and goals of therapy are discussed by Berry and Steiner in the Proceedings[636]. Nutritional requirements are reviewed by Leonard[637]. No controlled trial evidence for L-carnitine therapy in humans is available but eight OTC deficient patients had secondary carnitine deficiency proven during acute attacks and had fewer attacks when given oral supplements of 50–120 mg/kg per day of L-carnitine[638]. Protein requirements are adjusted for age in Leonard's Table 1[637], dropping from as high as 2.69 g/kg per day in the first month to 0.8 at age 18. Micronutrient deficiencies of iron and zinc may occur on a low protein diet. Plasma levels should be monitored and low values treated. Essential amino acid supplements are available for dietary help in severe OTC deficiencies[637].

Alternative pathway therapy history, rationale, pharmacokinetics in children, and adverse effects are discussed by Batshaw, MacArthur and Tuchman[639]. Goals of therapy are plasma ammonia concentrations less than twice normal, plasma glutamine levels $<1000\ \mu M$, normal growth and reduction in hyperammonemic crises to a minimum by aggressive home and if necessary hospital management[639].

If we knew how hyperammonemia damages the brain, perhaps we could design treatments that block its effects at the cellular level. The main theories include (1)increased tryptophan transport into the brain, leading to increased serotonin production and release, causing depression of cerebral function(736); (2) astrocytic synthesis of glutamine leading to brain swelling, which can be blocked by L-methionine S-sulfoximine(737); (3)increased extracellular brain glutamate concentrations that may cause seizures and cerebral edema(738); (4)altered protein phosphorylation of microtubule-associated protein, of Na+/K+-ATPase and NMDA receptors(739); (5)activation of the NMDA receptor-nitric oxide-cyclic GMP cycle which can be blocked by MK801(740); (6) activation of the peripheral-type benzodiazepine receptor on brain mitochondrial membranes which increases neurosteroid synthesis(741); (7) and impaired cerebral energy metabolism(742). Unfortunately, none of these theories or drugs effective in animals have so far led to novel treatment of hyperammonemia in humans.

In spite of optimal management, half of the children who survive neonatal onset OTC deficiency die by age six and those who survive have mild to severe developmental disabilities(522,639). However, long term therapy with low protein diets, arginine or citrulline and sodium phenylbutyrate at a dose of 350 mg/kg per day (range 125–484) in four late onset males and five affected females was successful in preventing hyperammonemic crises for a median of 26 months (range 17–42)(640). Protein intakes increased to 0.95 g/kg per day (range 0.66–1.46) while median plasma ammonia and glutamine were 30 and 902 μM, respectively.

Overall, symptomatic females(407,465,466), late onset males(406,504,511,530) and early onset males have done poorly even if they survived. Thus liver transplantation was adopted to treat OTC and other severe urea cycle deficiencies(641,642). We have referenced many of these transplant cases in our Tables 7-1, 2 and 3(433,436,438,484) and added other reports about the surgical techniques and problems(643–647). To summarize briefly, if a liver of proper size can be found, orthotopic total transplantation essentially cures the OTC deficiency but substitutes a life on immunosuppression with its risks of rejection, infections, drug nephrotoxicity and late malignancies(642). The scarcity of liver donors and the problem of sizing for small children led to use of split-liver cadaver transplants, replacing the left lobe and leaving the original right lobe as a reserve of other liver functions. The use of living-donor transplantation of a lobe or one of its segments has raised ethical questions because donors have died or suffered major complications. In OTC deficiency, citrulline remains deficient because of OTC lacking in the gut and arginine therefore becomes deficient, requiring supplementation[644]. Overall patient survival rates for one year are 92%, and 82% for five years while graft survivals are 85% and 71% respectively[642]. The most important lesson learned is that no improvement in developmental defects takes place *after* transplantation so patients at risk for brain damage but still little damaged neurologically should be transplanted *before* any further injury can occur, even including severely affected female heterozygotes[648].

Another issue in the treatment of OTC deficiency is the need to help families cope with the stresses that this complex illness imposes on the affected person and caregivers. A survey of families who have a child with a urea cycle defect found that the three greatest stressors were financial problems, fear of the child's death and restrictions imposed by the diet[649]. Readers who care for such children and their families should read the article by J. A. Cederbaum et al.[649] for the details of their findings. The study proves to us that an inter-disciplinary team of geneticist-physicians, nephrologists, nurses, social workers, psychologists and financial/insurance advisors is needed to care for such a complex illness as OTC deficiency.

Gene Replacement Therapy

Germ-line correction of OTC deficiency, though occasionally successful in *spf* mice (Chapter 6), is not predictably successful enough to warrant trials in humans. Gene insertion into human ova raises many ethical questions as well. Therefore somatic gene therapy has been the main effort of those focusing on correcting OTC deficiency. The development of adenovirus vectors for temporary correction of the deficiency in the spf mouse and the problems encountered in this species are described in Chapter 6. The review by Lee and Goss[642] describes the problems and ultimate failure of the one human protocol to treat OTC deficiency with an E1/E4-depleted adenovirus vector expressing human OTC[650]. The protocol was not well followed in the last patient, an 18y old male with late-onset disease, who died of a generalized cytokine and immune response to the vector, forcing the Food and Drug Administration to put a "clinical hold" on enrolling any new patients[651,652] and a civil lawsuit has been brought against all involved[744]. Because of the immune response to any viral capsid proteins and low levels in the liver of transfected OTC cDNA, the therapeutic index has remained quite narrow. Among the 18 patients who tolerated increasing doses of the vector, side effects of fever, chills, myalgias, nausea and vomiting always occured[745]. None of seven liver biopsies showed more than 1% of hepatocytes with transgene expression. A drop in serum phosphate and platelet counts was noted. Neutralizing antibodies developed in all subjects[745]. Overall, this attempt at adenoviral vector therapy of partial OTC deficiency was a great disappointment.

Development of adenoviral vectors devoid of all viral genes, and persistent transgene expression of many kinds in animals has been described[642]. Adenovirus-associated vectors are an alternative to regular adenoviruses and have the advantages of less vector-associated toxicity and also yield genomic integration that gives prolonged transgene expression. However problems remain to be solved with this vector. Strategies to avoid immune clearance of hepatocytes containing the transgene are discussed[642]. Perhaps the "gutted" adenoviral vectors containing DNA transposons will be effective in OTC therapy (see Chapter 6)[723,724]. Further attempts to correct OTC deficiency by somatic gene therapy should await successful transgene expression in primates before careful attempts in humans are tried again.

Commentary. The diagnosis of OTC deficiency is now straightforward and possible in most medical centers but the diagnosis *must* be suspected and the plasma ammonia must be measured. The detection of female carriers or of affected males with minimal symptoms has been made safe and efficient using the allopurinol loading test while the subject is consuming a high-protein diet. When the allopurinol test is negative but the pre-test likelihood is high, DNA analysis of all 10 exons can be done at referral centers. Treatment must be aggressive early in hyperammonemic episodes to prevent brain damage. When diet, alternative pathways for nitrogen excretion and arginine or citrulline treatments cannot prevent relapses or brain damage, prompt liver transplantation is indicated. Somatic gene therapy looks promising but has yet to come to fruition.

9

Induction and Suppression of OTC and Urea Cycle Enzymes in Bacteria, Fungi and Mammals

Human OTC is related phylogenetically to the OTCs of the rat and mouse, less so to that of the frog and distantly to the six fungal OTCs which have been sequenced, called *OTCβ*s by Labedan[172]. The OTCs of *E.coli* and *P.aeruginosa,* which like hOTC have had their structure determined by X-ray diffraction, belong to the *OTCα* group by phylogenetic analysis of their amino acid sequences[172].

The control of amounts of OTC enzyme protein differs in bacteria from controls in fungi and from those in mammals, although some mechanisms have carried over to hOTC in its induction and suppression in hepatocytes. The arginine biosynthetic pathway in *E.coli* consists of nine enzymes and 10 genes, one coding for the argR repressor protein.[662] OTC is the sixth enzyme in the *E.coli* pathway from glutamate to arginine. The ninth enzyme is a glutamine-dependent carbamate kinase that supplies CP for OTC and for pyrimidine synthesis. Arginine exerts feedback inhibition of the first enzyme and also binds to the arginine repressor protein preventing transcription of the eight genes[662]. Thus in *E.coli* and most bacteria, induction of OTC is accomplished by *de-repression* when arginine levels are low. An excellent summary of the development of de-repression in the arginine pathway is that of Maas[663].

Among eukaryotic fungi, the cellular compartmentation of *Neurospora crassa* most resembles that in mammalian hepatocytes (Figure 9-1)[664]. The 5 enzymes that convert glutamate to ornithine, and enzyme 6 (OTC) and enzyme 9 (CPS-A) are all mitochondrial. Enzyme 7 (AS) and enzyme 8 (AL) and the newly evolved enzyme in fungi, arginase (enzyme 14), are cytoplasmic. Arginase functions as a catabolic enzyme along with urease (enzyme 18a) and ornithine transaminase (enzyme 15). Fungi contain vacuoles that store arginine and ornithine and contain hydrolases as do hepatocyte lysosomes. When *N.crassa* cells are replete with arginine, it enters the mitochondria and by feedback inhibition shuts down enzymes 1 and 2. Arginine also blocks the uptake of ornithine into mitochondria. Thus OTC has little ornithine to make unneeded citrulline. Arginase and ornithine transaminase are induced by excess arginine. Urea generated by arginase yields ammonia from urease action, which is used as a nitrogen source. The excessive ornithine is converted to glutamate semialdehyde (enzyme 15) which cyclizes to Δ^1-pyrroline-5-carboxylate (P5C) that is converted to proline by enzyme 12, P5C reductase (Fig. 9-1).

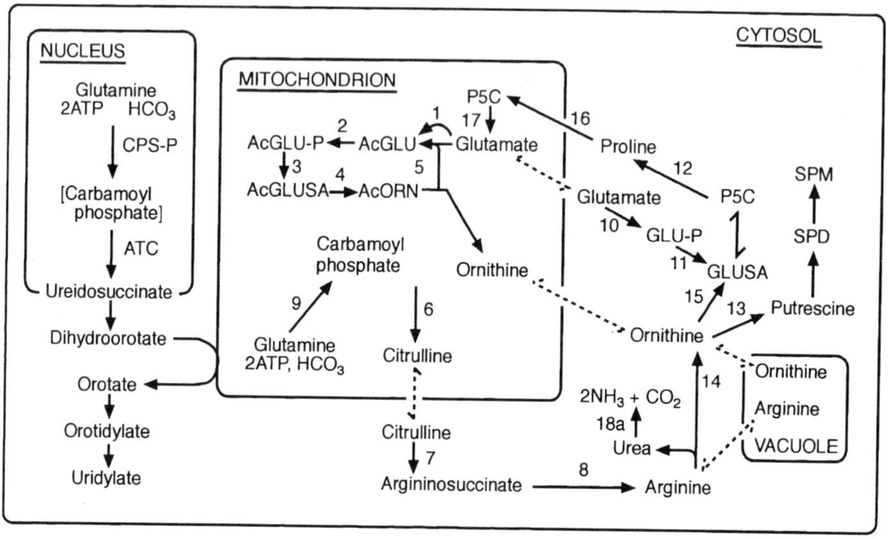

Figure 9-1. Organization of arginine, pyrimidine, proline and polyamine metabolism in *Neurospora crassa*[664]. Abbreviations: ATC,aspartate transcarbamylase; CPS-P, pyrimidine-specific carbamyl phosphate synthetase; Ac-,acetyl; GLU, glutamate; GLUSA, glutamate semialdehyde; ORN, ornithine; P5C, pyrroline-5-carboxylate; SPD,spermidine; SPM, spermine. (Copyright 1986. Reprinted with permission from Amer Soc Microbiol.)

When arginine is severely limited in the growth medium, *N.crassa* has three means for generating it within the cell[664]. The first means is by drawing on storage from the vacuoles, releasing arginine and ornithine into the cytoplasm. The second means, derepression of arginine biosynthetic enzymes, in *N.crassa* requires an arginine concentration one quarter of that in a minimum medium where anabolic and catabolic processes are in balance[664]. The third means is called the *general amino acid control* system, regulated by GCN genes 1–4[664]. Deficiency of one amino acid derepresses the synthetic pathways of many other amino acids. In the arginine pathway of N.crassa, arginine deficiency, or that of leucine for example, derepresses synthesis of enzymes 1–9 (Fig. 9-1). The general amino acid control system is the major mechanism for correcting arginine or other amino acid deficiencies in N.crassa[664]. Thus these complex controls in fungi shut off wasteful urea formation when arginine is in short supply, stop arginine synthesis when it is plentiful and channel excess arginine to ammonia, ornithine, proline and glutamate for growth and energy.

The mammalian urea cycle is shown on Figure 9-2 along with the dibasic amino acid transporters that coordinate with the cycle. Also included in Fig. 9-2 are the inherited urea cycle defects. The nomenclature used in Fig. 9-2 is summarized in Table 9-1 which includes the deficiencies of 5 urea cycle enzymes and that of N-acetylglutamate synthase and two transporter defects, as well as the tissue distribution of the enzymes, the chromosomal localization of their genes, and their MIM numbers[665]. Genetic defects have been described for the ornithine transporter in the mitochondrial membrane of the hepatocyte and the dibasic amino acid transporter in the plasma membrane.

In the mammalian urea/ornithine cycle, an ammonia and acetylglutamate-dependent CPS-1 is located in the mitochondria, serving for citrulline synthesis. A glutamine-dependent CPS-2 is found in the cytoplasm, supplying CP for pyrimidine synthesis[666]. As in bacteria and fungi, CPS-2 is inhibited by nucleoside polyphosphates, especially UTP, which inhibits and changes the kinetics with ATP from a slightly sigmoidal curve to a strongly sigmoidal curve. CPS-1 is not inhibited by UTP. Arginine does not cause feedback inhibition of CPS-1 or OTC but *stimulates* the activity of acetylglutamate synthetase[666]. In bacteria and fungi, arginine inhibits conversion of glutamate to acetylglutamate. Overflow of CP from mitochondria to the cytoplasm occurs in OTC deficiency and in any condition which reduces mito-

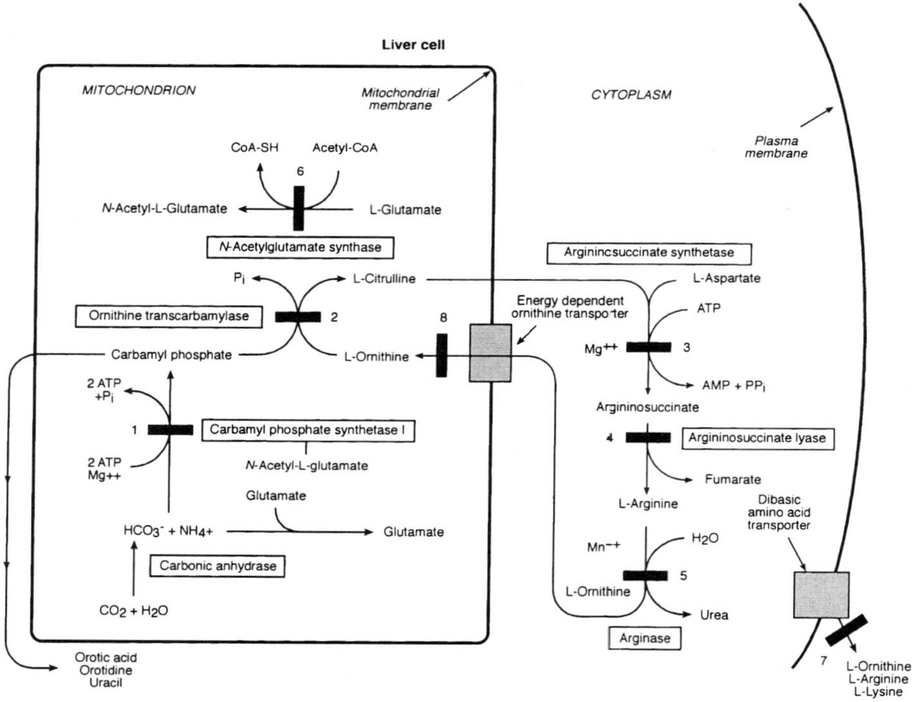

Figure 9-2. Urea cycle and associated enzymes in the human liver cell, and their inherited defects(665). Numbering of the enzymes is shown in Table 9–1. (Copyright 1996. Reprinted with permission from Oxford University Press.)

Table 9-1. Nomenclature for Figure 9-2

No.	Disorder-affected component	Tissue distribution	Chromosomal localization	MIM no.
Urea cycle enzymes				
1	Carbamyl phosphate synthetase I (CPS)	Liver, gut (mitochondria)	2p	237300
2	Ornithine transcarbamylase (OTC)	Liver, gut, kidney mitochondria)	Xp 21.1	311250
3	Argininosuccinate synthetase (citrullinemia, CITR)	Liver, kidney fibroblasts (cytosol)	9q34	215700
4	Argininosuccinate lyase (arginosuccinic aciduria, ASA)	Liver, kidney fibroblasts RBC, brain (cytosol)	7cen-q11.2	
5	Arginase (argininemia, ARG)	Liver, kidney, RBC (cytosol)	6q23	207800
6	N-Acetylglutamate synthase (NAGS)	Liver, gut, (mitrochondria)	Unknown	237310
Transport defects of dibasic amino acids				
7	Lysinuric protein intolerance (LPI) (hyperdibasic aminoaciduria)	Liver, gut, kidney (plasma membrane)	Unknown	222700
8	Hyperornithinemia-hyperammonemia-homocitrullinuria (HHH)	Liver, fibroblasts (mitochondrial membrane)	Unknown	238970

chondrial ornithine levels, causing excessive synthesis of orotic acid and orotidine, as discussed in Chapter 8.

The major carry-over of a control mechanism for enzyme synthesis from the bacterial and fungal arginine synthetic pathways to mammalian hepatocytes is the derepression by arginine deficiency of CPS-1, OTC, argininosuccinate synthetase (AS) and lyase (AL). Schimke fed weanling male Osborne-Mendel rats a low arginine diet and "de-repressed" or induced these four enzymes and suppressed arginase levels(26). We confirmed the induction of the first four enzymes in Wistar male rats and partial responses in 3 other rat strains and in a mouse(27) but did not obtain a decrease in arginase by any baseline. This induction occurred even though liver concentrations of arg, orn, citrulline and urea were slightly but not significantly lower on arginine-free diets(27). Synthesis of arginine from glutamine in the intestine/kidney axis(746) maintains arginine levels. In fungi the general amino acid control system turns on synthesis of arginine even though the cytosol level of arg is only slightly low: the postulated signal is the accumulation of uncharged tRNA(664).

The same phenomenon was shown more impressively by Schimke for AS and AL in HeLa cells and in two fibroblast lines in culture(28), where arginine concentrations in the growth media were widely varied. Arginase levels were suppressed by low arginine media and increased by high arginine media, as in fungi(28,664). Arginine–mediated repression of human AS was demonstrated in RPMI 2650 cells(25). Repression of AS occurred in these cells when cultured in citrulline-containing medium lacking arginine(668). Morever expression of a transfected minigene containing the 5′-flanking region of AS was increased by *leucine* starvation, due to increased rates of transcription. However, AS enzyme activity only increased 2-fold in 72h of arginine starvation and required 3 weeks to show maximal derepression(667). This leucine effect resembles that of the general amino acid control system in fungi(664) where starvation for leucine derepresses its synthetic pathway and that of other unrelated amino acids, here the arginine pathway. Bruhat et al.(668,669) have found that leucine deprivation and that of arginine and other amino acids in HeLa cells and in HepG2 cells (an hepatocyte-derived tumor cell line) induces AS and a CCAAT/enhancer-binding protein-related gene (CHOP), because of a cis-positive element in the promoter region of the human CHOP gene. This element contains a minimal core sequence 5′-

ATTGCATCA-3′ which has been named an amino acid response element (AARE). AARE binds CHOP and activating transcription factor 2 (ATF 2). CHOP protein-C/EBP heterodimers play a role in regulation of stress-induced cell responses concerning energy metabolism, cell proliferation and expression of cell type-specific genes[669]. Urea cycle enzyme genes eventually may be shown to be derepressed by CHOP-C/EBP binding to elements of their promoters when arginine starvation occurs. However, we could not identify the core AARE sequence on the promoter region out to −578bp of the hOTC gene[10] or −800bp of the mouse gene[15] or −1011bp of the rat OTC gene[16].

We were unable to obtain any de-repression-like responses to media lacking arginine with rat liver cells cultured in monolayer for 72h or in a cultured hepatoma line maintaining all five urea cycle enzymes[27]. The arg content of cells cultured in Minimal Essential Medium (MEM) without arginine was undetectable. These cells could still be induced by glucagon and dexamethasone as well as cells in arg-containing MEM. Perhaps 72h was too short a culture time to see derepression of urea cycle *activities* in arg-free MEM[667].

The amounts of the 5 mammalian urea cycle enzymes in liver are related to the level of *protein intake.* In 1962, Schimke showed in rats that the activities of all 5 enzymes were proportional to the level of dietary casein intake[341] and that the amounts of OTC and arginase were increased as well. The induction by the 60% casein diet peaked at 8 days for OTC and arginase. Addition of any one of the intermediates of the pathway (ornithine, citrulline, arginine or urea) to a low casein diet did not alter enzyme levels[26]. No induction occurred when the amounts of the following amino acids found in a 60% casein diet were added singly to a 15% casein diet: histidine, tryptophan, aspartate, methionine, alanine, threonine, lysine and leucine. Degradation as well as enzyme synthesis played a role in arginase levels[670]. Using ^{14}C-guanidinoarginine that is not re-utilized, the transition from a high to a low protein diet resulted in suppression of synthesis and a marked increase in arginase degradation until a new steady state was reached. During starvation, which is equivalent to a high protein diet because endogenous proteins are utilized, degradation ceased and synthesis continued at a constant rate until a higher steady state of arginase protein was achieved[671]. The half-life of arginase was 4–5 days by this method. It is remarkable that during starvation the urea cycle enzymes are increased in amount while the bulk of hepatocyte proteins are degraded, resulting in a smaller liver per body weight and a lowered total protein concentration[671].

We found that we could induce all 5 urea cycle enzymes in 48h when male Sprague-Dawley rats stabilized on a 15% casein diet were tube-fed casein hydrolysate three times a day at a dose of 315 mgN (2 gN/kg/day)[342]. OTC increased 1.4-fold expressed as units/liver per 100g of rat; the increments of the other enzymes were CPS 1.5-, AS 1.9-, AL 1.7- and arginase 1.3-fold. The casein feedings also increased liver weight per 100g of rat and liver protein concentration. In enzymes like arginase where the amount of enzyme protein in liver is large and the rate of degradation on an adequate (15%) protein intake is slow, even a 4-fold increase in synthesis rate may lead to what appears to be small increases in activity per liver[670]. We believe this same interpretation explains the small increases in OTC.

This two-day tube-feeding regimen was then applied to rats fed a 15% casein diet by giving them maximally tolerated doses of individual amino acids, avoiding toxicity[342]. Alanine (2gN/kg), glycine (2gN/kg), methionine (0.2gN/kg) and cysteine (0.4gN/kg) each increased all 5 enzymes. A mixture of 8mgN of met, 12mgN of ala and 16mgN of gly, the amount of these amino acids present in 315mgN of casein hydrolysate, induced as well as the mixed amino acids of this dose of casein. Arginase increases were as great by immunoprecipitable protein as by activity. OTC was the least inducible of the five enzymes, ranging from 1.2-fold by cys to 1.5-fold by met. Tube feedings of pyruvate (85mmoles/kg) were compared with the same sized dose of ala (1.2gN/kg) but only ala induced all 5 enzymes (OTC 1.4-fold). The α-OH-acid of methionine (17 mmoles/kg) was compared with an equal dose of met but only the amino acid induced all 5 enzymes (OTC 1.3-fold). Thus the inducing effect of amino acids requires the presence of the α-amino group. Our success with induction by amino acids that did not induce in Schimke's study we attribute to our higher doses, for example, alanine[670]. The explanation for the induction by ala, gly, met and cys plus some contributions by other amino acids[342] is not well understood but implies that there are positive amino acid response elements in the promoter regions of these five urea cycle enzyme genes, on five different chromosomes. One of these elements may be the urea cycle response element found in the 5′-region of OTC, AS, arginase and ornithine transaminase[17,18] which may control coordinate induction by dietary protein.

Pitot and Peraino found that tube-feeding casein hydrolysate to rats induced hepatic threonine dehydratase and ornithine transaminase in 24h (the method we adopted)[672]. When glucose was fed along with the casein the induction was blocked totally[673]. Was this a carry-over of glucose repression in bacteria into mammals? We looked for this phenomenon in the urea cycle by stabilizing rats on an 8% casein diet, inducing all five enzymes with casein hydrolysate (3.9g/day) and comparing the responses to those when 3.9g/day of glucose was fed along with the casein. CPS, OTC and arginase induced as well as with casein alone but AS was suppressed by 20% and AL by 21%, a barely significant non-coordinate suppression[674]. During glucose repression of casein induction of threonine dehydratase, hepatocyte cAMP levels were not reduced as occurs in glucose repression in bacteria[675,676]. Glucose fed to starved rats elevated serum insulin while suppressing glucagon, and hepatic cAMP levels were markedly reduced[677]. When casein hydrolysate was fed, both insulin and glucagon levels rose but the cAMP level in liver fell 25% and returned to baseline.

To test whether ammonium ion is an inducer, we stabilized rats on a 27% casein diet and tube-fed ammonium citrate (2.2gN/kg), a maximum tolerated dose, three times daily for 8 days[342]. Only AS increased, 1.2-fold. The urea excretion was the same as that of rats fed casein hydrolysate (2gN/kg) that induced all 5 enzymes. Thus processing ammonia through the cycle was not the signal for urea cycle induction.

A major evolutionary addition to the induction and suppression of animal enzymes is the role of *hormones*. We tested glucagon as an inducer of the 5 urea cycle enzymes because pharmacologic doses of soluble glucagon given to rats had been shown to increase CPS-1, AS and AL but not OTC or arginase[678]. This report also showed that alloxan diabetes increased all five enzymes, and that insulin (4 units/day of protamine Zn insulin) did not affect any enzyme level in normal rats and did not restore enzyme levels to normal in diabetic rats. We used Zn glucagon because it has a prolonged action when given subcutaneously. When we gave *pharmacologic* doses of Zn glucagon to male rats, in 48h we increased CPS 2-fold, OTC 1.3-fold, AS 2.7-fold, AL 3.2-fold and arginase 2.2-fold[679]. *Physiologic* doses by an osmotic pump for 7 days induced all 5 enzymes: CPS 1.5-fold, OTC 1.2-fold, AS 2.7-fold, AL 1.8-fold and arginase 1.6-fold. Actinomycin-D or puromycin or cycloheximide blocked these increases, indicating that

RNA and protein synthesis are necessary for the induction. Tube-feeding casein hydrolysate for 2 days induced the entire cycle while causing an increase in plasma glucagon equal to that of our physiologic dose of glucagons(679).

Schimke reported that adrenalectomy reduced all 5 urea cycle enzymes(26) and McLean and Gurney confirmed this(680). We found that adrenalectomy reduced all enzymes but AL when urea cycle activities were expressed as units/liver per 100g of rat(343), a baseline which adjusts for changes in liver and body weight. In this report we gave continual high dose glucagon (0.3 mg/kg/day) to intact rats for 7 days and induced all enzymes but OTC. The only enzyme induced by glucagon in adrenalectomized rats was AL. Physiologic doses of dexamethasone induced arginase alone in intact and adrenalectomized rats. The combination of the two hormones induced all 5 enzymes synergistically in intact rats but the induction was additive or less than additive in adrenalectomy rats(343). What role other hormones might play is not clear. Thyroidectomy increased and excess thyroid hormone decreased OTC, CPS, AS and AL but arginase was decreased in both conditions(681).

We did not obtain a good correlation between portal vein plasma glucagon levels after tube-feeding various amino acids and their ability to induce the urea cycle, so we did not conclude that protein feeding induced the cycle only by release of glucagons(674). Eisenstein(682) found that high protein feedings elevated portal vein plasma glucagon levels but there was no correlation between glucagon and amino acid concentrations in the portal vein. He also studied the release of glucagon when rat islets were perifused with various amino acids: arginine and glutamine and less so ornithine were the major releasors(683). Ala, cys, gly and met, the amino acids we found induced the cycle(342), did not release glucagon well. In dogs, plasma glucagon levels were measured after amino acid infusions(684). The potencies of releasors were as follows: asn > gly > phe > ser > asp > cys > try > ala > glu > thr > gln > arg > orn > pro > met > lys > his. Val, leu and ile failed to stimulate glucagon release(684). Thus the induction of urea cycle enzymes by amino acids or by glucagon in whole rats seems to be by separate mechanisms and is not simply caused by amino acids releasing glucagon. To illustrate the complex interactions, dexamethasone injection in humans caused basal plasma glucagon and alanine levels to rise and alanine infusions or protein ingestion caused greater glucagon responses(685).

Mori et al.(686) first showed that the liver translatable mRNA levels in rats fed a 5% casein diet increased 4.2-fold for CPS and 2.2-fold for OTC when the rats were fed a 60% casein diet for 8 days. The mRNA increases were greater than the enzyme activity increases. Starvation for 7 days increased CPS and OTC activities but decreased mRNA levels by 40–50% of control animals, suggesting a decreased rate of protein degradation. Morris et al.(687) further showed that hybridizable levels of mRNA for all 5 enzymes in rat liver were proportional to dietary protein levels: OTC mRNA increased 4.4-fold 6 days after a dietary increase from a 27% to a 60% casein diet. Contrary to Mori(686), Morris reported a marked increase of mRNAs for all 5 enzymes after 5 days of starvation(687).

Controlled conditions are difficult to maintain in studies of whole animals. Results of enzyme or mRNA changes may be due to indirect effects of dietary changes(343,678–681,686–687) and various hormonal releases by amino acids add to the interpretive difficulties(682–685). We therefore developed a method to culture normal adult rat liver cells which maintained urea cycle enzymes in chemically-defined, serum-free medium for at least 7 days, using calf skin collagen to coat the plates(688). When the hepatocytes were cultured in 10% fetal calf serum during the first 21h the polygonal cells formed a monolayer which then stayed well attached in serum-free medium for 4d, long enough to do urea cycle induction studies. Adenylyl cyclase activities were stable and inducible for 72h. We used this cell culture system successfully from 1975 to 1993.

Gebhardt and Mecke used rat liver cells cultured in perifused monolayers and showed that 1μM glucagon or 20μM dibutyryl-cAMP induced all urea cycle enzymes but OTC(689). Dexamethasone (0.1μM) induced only AS and arginase. The combination of glucagon and dexamethasone induced all five enzymes in 24h in a manner called a permissive effect of dexamethasone, OTC increasing least at 1.3-fold(689).

In our monolayer culture system we added bacitracin to protect glucagon from degradation and were able to increase cAMP levels for 7h or more after adding glucagon(690). We used physiologic concentrations of glucagon (1nM) added 4 times in 48h and induced only CPS-

1. When 10nM dexamethasone was added twice in 48h to the glucagon regimen, all five enzymes were induced: CPS 1.5-fold, OTC 1.2-fold, AS 1.9-fold, AL 1.8-fold and arginase 1.7-fold. Dexamethasone alone induced only AS and AL. Amounts of immunoprecipitable AL and arginase were increased as much as their activities. Glucagon induction was not affected by added insulin. Other hormones that did *not* affect any urea cycle activity with or without dexamethasone included epinephrine, pentagastrin, secretin, cholecystokinin octapeptide, dimethyl prostaglandin E2, triiodothyronine, estradiol and growth hormone(690). We concluded that glucagon, via cAMP, supported by a permissive role of glucocorticoids, is the only identified coordinate hormonal inducer of all five urea cycle enzymes.

Morris extended our understanding of hormonal induction by showing that intraperitoneal injection of dibutyryl cAMP (dbcAMP) in whole rats increased 4 urea cycle mRNAs at 5h but not that of OTC(687). Transcriptional run-on assays showed 4–5-fold increases in CPS amd AS in 30 min. In cultured rat hepatocytes Nebes and Morris found that in 16h CPS, AL and arginase mRNAs increased during exposure to 8-(4-chlorophenylthio)cAMP (CPT-cAMP)(691). OTC mRNA did not respond. Dexamethasone increased CPS, AL and arginase mRNAs more slowly than with CPT-cAMP. Dexamethasone but not CPT-cAMP inductions were blocked by cycloheximide, suggesting that dexamethasone induced another gene product which led to urea cycle mRNA induction. The combination of both hormones gave synergistic mRNA responses for CPS, AS and arginase, additive responses for AL and *no* response for OTC. The hepatic mitochondrial ornithine/citrulline transporter was also induced by increased dietary protein(692). Moreover its mRNA was induced to a maximum in 12h by CPT-cAMP and in 4h by dexamethasone and the combination was synergistic, rising 34-fold after 16h of treatment.

Our own studies on mRNA induction by glucagon and dexamethasone demonstrated that the coordinate induction is accomplished by three different mechanisms(693). The transcription rate for arginase mRNA increased 9-fold in 7h, the mRNA 90-fold in 28h and the activity 1.5-fold at 48h when 1μM glucagon and 10nM dexamethasone were added to cultured rat liver cells. Arginase mRNA induction was minimal with either hormone alone. Cycloheximide pretreatment did not prevent the rise in mRNA levels with both hormones. CPS-1 mRNA levels also responded synergistically to both hormones, peak-ing 240-fold above controls at 24h although activity only increased 1.4-fold at 48h. No transcriptional assays were done for CPS-1, but we assumed that transcription rate increases would not approach the 240-fold increase in mRNA levels. We interpreted these arginase responses to result from stabilization of mRNA by the two hormones because the steady-state mRNA increase was so much greater than the increase in transcription rate. AL and AS mRNAs were not induced by single hormones but responded to the combination by a 10-fold increase in AL and a 7-fold rise in AS transcription rates at 7h. AL mRNA peaked at 28h 7-fold above controls while AS mRNA peaked at 52h 14-fold above preinduction levels. AL enzyme activity increased 2.8-fold at 48h and AS 1.8-fold. We interpreted the induction mechanism for AS and AL as transcriptional control primarily. OTC mRNA levels were not increased by the combined hormones yet the activity increased 1.3-fold. We interpreted this result as consistent with OTC protein stabilization(693).

Matsuno et al.(694) found that dexamethasone and glucagon rapidly increased the transcription factor CCAAT/enhancer binding protein β (C/EBPβ) mRNA synergistically. The transcription rate and the DNA-binding activity of C/EBPβ were induced as much as the mRNA. C/EBPα mRNA did not respond. C/EBP family members are found in the CPS-1 promoter, in the arginase promoter and enhancer and in the OTC enhancer where they work in combination with hepatocyte nuclear factor-α (HNFα)(694). These factors may play a role in secondary glucocorticoid responses where the time course of induction is delayed and is sensitive to protein synthesis inhibitors(691).

Kimura et al.(698) made *C/EBPα* knockout mice and found decreased levels of mRNA for CPS, AS, AL and arginase in newborn or 5–7h mice while OTC mRNA was variably low. Protein levels by Western blots were low for the same four enzymes whereas OTC was reduced by 60% and its lobular distribution was shifted to the mid-lobular from the usual peri-portal region. Kimura further showed that *C/EBPβ* knockout mice still showed a normal *basal* level of mRNA and protein for all 5 enzymes(696). When glucagon and dexamethasone were given intraperitoneally to wild-type mice, the liver CPS, AS, AL and arginase mRNAs were induced normally but OTC did not increase. In cultured mouse hepatocytes the wild-type urea cycle responses to the combined hormones were lost. OTC showed no induction in wild-type or C/EBPβ deficient cells(696). Apparently the deficiency of C/EBPβ was compensated for *in vivo* by other transcription factors.

One new finding for gene regulation of urea cycle enzymes was that induction of CPS-1 and AS mRNAs by dexamethasone was blocked by long-chain fatty acids, arachidonic acid being the most potent[697]. No inhibition of induction by dbcAMP was seen with oleic acid. This finding may help explain the suppression of urea cycle enzyme activities and the hyperammonemia in hereditary carnitine-deficient JVS mice whose serum free fatty acid levels are very high[697].

Based on the findings reported in references 20–22 concerning factors which bind to regulatory regions of the OTC gene and from expression of OTC in liver vs. intestine in transgenic mice[346–348], Takiguchi and Mori have constructed models of the OTC enhancer and promoter binding of the liver selective factors C/EBP and HNF-4α (Figure 9-3)[698]. They have illustrated in Figure 9-4 how these factors cooperate in the OTC enhancer to make transcription liver-pref-

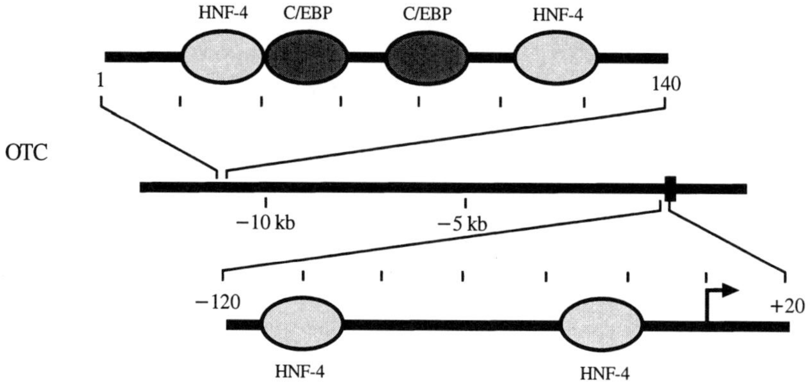

Figure 9-3. "Role and regulation of the OTC promoter and enhancer in tissue-selective transcription. In transgenic mice the OTC promoter exhibits higher activity in the small intestine than in the liver; the enhancer inverts this tissue–selectivity of the promoter, bringing about higher expression in the liver. The OTC promoter is activated by HNF-4, a liver- and small-intestine selective member of the steroid receptor superfamily and is competitively repressed by COUP-TF, a ubiquitous member of this family. For activation of the enhancer, the two liver-selective factors HNF-4 and C/EBPβ are required and neither alone is sufficient. This appears to enable these liver-selective, but not strictly liver-specific, factors to confer more restricted liver-specificity on expression of the target OTC gene." Legend quoted from Takiguchi and Mori[698].(Copyright 1995. Reproduced with permission from the Biochemical Society.)

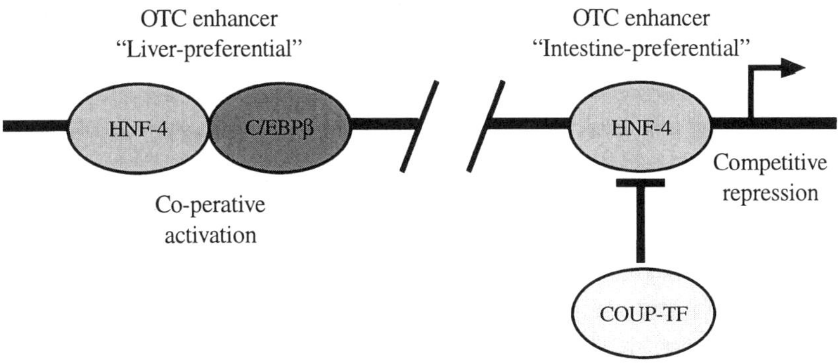

Figure 9-4. "Factors binding to regulatory regions of the OTC gene. The transcription start site is indicated by the "hooked' arrow. The enbancer regions were identified 11 kb upstream from the start site of the OTC gene. C/EBP represents several members of the C/EBP family. HNF-4 sites can be recognized also by COUP-TF and possibly by other factors." Legend quoted from Takiguchi and Mori[698]. (Copyright 1995. Reproduced with permission from the Biochemical Society.)

erential. This contrasts with the binding of HNF-4α to the OTC promoter that allows intestinal expression because this expression can be competitively repressed by binding COUP-TF. Binding of factors to enhancers or promoters of the other 4 urea cycle genes is illustrated and discussed in reference 698. Mice.with a liver-specific disruption of the HNF4α gene showed a marked *decrease* in liver OTC mRNA, a 60% decrease in argininosuccinate lyase mRNA but an increase in mRNAs for CPS-1, AS and arginase, resulting in increased serum ammonia and urea levels[748]. One other transcription factor, peroxisome proliferator-activated receptor α, inhibits mRNA expression for all five urea cycle enzymes in liver[749]. Morris has published an excellent review of the regulation of the urea cycle enzymes and has expanded the discussion to include regulation of arginine synthesis by the intestine/kidney axis, the citrulline-nitric oxide cycle and arginine catabolism[746].

Development of Fetal and Neonatal OTC and Urea Cycle Enzymes

Development and control of the urea cycle enzymes in liver have been studied most thoroughly in the rat. We will summarize briefly the results of many reports, including what is known about the mouse and human urea cycle development.

OTC activity or protein becomes detectable in the rat fetus on the fourth day before birth (−5d) or on day 17 of gestation, which is 21.5–22.5 days in the rat[700]. OTC activity rises rapidly to a peak at 4d, then declines and rises slowly to a plateau from +20 to +35 days[700], equal to adult levels[699,702]. During days −4 to birth, the mass of liver hepatocytes doubles while the hematopoietic cells recede[703]. From −8 to −5 days of fetal life only 18% of liver consists of hepatocytes. The OTC gene is substantially methylated at the 5′-end of the gene at −15d and demethylation of one OTC allele occurs in liver but not kidney in the next few days[703]. Thus the rise in activity is due to proliferation of hepatocytes, hypo-methylation of the 5′-end of OTC and new OTC synthesis. OTC mRNA in the rat liver becomes detectable at day −7, at 30% of adult levels[703]. The mRNA rises steadily to birth, when a brief drop occurs and then rises steadily to adult levels by day +7. This pattern is found similarly in CPS−1 and arginase[703], the so-called late fetal type of maturation.

Hormonal changes in the fetus seem to play a role in this early fetal enzyme development. Corticosteroids are elevated at day −1, rise to a peak at −2d to birth, then fall and rise again from +9 to +16d[701]. Insulin levels in the fetus are high days −2 to birth when they decrease to low levels and rise again after weaning. Glucagon levels are low before birth, rise sharply at birth, and then decline to low levels. Cyclic AMP levels follow the glucagon pattern[701]. OTC, CPS and arginase are increased at +13d by exogenous prednisolone, by glucagon and by the combination in a less than additive manner[701]. AS mRNA was detected at 15.5 days (−7d) in the fetal rat liver and increased steadily to day +2, along with AS activity, due to an increased transcription rate[706]. In cultured fetal hepatocytes dexamethasone increased AS mRNA, glucagon enhanced the dexamethasone effect and insulin suppressed it[706]. AS activities in the rat liver were not reported ante- or post-natally.

In the spiny mouse the activities of OTC, CPS and arginase activities increased from −5d steadily to adult levels at +35d[702]. Morris[704] found in Swiss-Webster mice that the mRNA levels for all 5 urea cycle enzymes were present at 2–14% of adult liver levels at day −6. OTC mRNA increased rapidly at days +1 to +4. AL mRNA rose from −2d to +2d. CPS, AS and arginase mRNAs increased rapidly from −4d to −2d and then declined slightly.

Human OTC activity and that of the other urea cycle enzymes were detected at 8 weeks of gestation, increased steadily to birth and up to 8 weeks of age(707). More extensive studies of human fetal livers from 13 to 36 weeks of gestation(705) showed that all five urea cycle activities were present at 13–16 weeks. OTC was then at 22% of adult liver levels, arginase at 28%, AL at 40%, CPS at 52% and AS at 55% of adult levels. By term human liver activities of all five enzymes were 65–90% of those in adult human liver.

Commentary. The human urea/ornithine cycle is an evolutionary descendant of the arginine synthetic and degradative pathways of fungi, and distantly of the arginine biosynthetic pathway of bacteria. The amounts of OTC protein, of CPS-1, AS, AL and arginase proteins, of N-acetylglutamate synthetase protein and of the ornithine/citrulline transporter in liver cells usually increase or decrease in a coordinated manner. Two gene control mechanisms for OTC and other urea cycle enzymes in fungi have carried over into mammals, namely arginine-deficiency derepression and general amino acid control. The amounts of OTC and other urea cycle proteins adapt to the level of protein intake, increase with starvation and also increase with secretion of glucagon and glucocorticoids. Coordinate induction of the urea cycle by glucagon and dexamethasone is accomplished by three mechanisms: in CPS-1 and arginase by stabilization of mRNAs, in AS and AL by increased transcription rates and in OTC by protein stabilization. The transcription factor, HNF4α, is essential for OTC gene expression in liver while another factor, peroxisome proliferator-activated receptor α, inhibits mRNA expression of all five urea cycle enzymes. Development of OTC and the other urea cycle enzymes in fetal and post-natal livers is complex and modulated by glucagon, corticosteroids and insulin.

References

1. Gene Structure, Regulation and Function

1. Lindgren V, DeMartinville B, Horwich AL, Rosenberg LE, Francke U. Human ornithine transcarbamylase locus mapped to band Xp21.1 near the Duchenne muscular dystrophy locus. Science 1984; 226:698–700.
2. Lyon MF. Gene action in the X-chromosome of the mouse (Mus musculus L.). Nature 1961; 190:372–373.
3. Ricciuti FC, Gelehrter TD, Rosenberg LE. X-chromosome inactivation in human liver: confirmation of X-linkage of ornithine transcarbamylase. Am J Hum Genet 1976; 28:332–338.
4. Horwich AL, Kraus JP, Williams K, Kalousek F, Konigsberg W, Rosenberg LE. Molecular cloning of the cDNA for rat ornithine transcarbamylase. Proc Natl Acad Sci USA 1983; 80:4258–4262.
5. Kraus JP, Rosenberg LE. Purification of low-abundance messenger RNAs from rat liver by polysome immunoabsorption. Proc Natl Acad Sci USA 1982; 79:4015–4019.
6. Horwich AL, Fenton WA, Williams KR, Kalousek F, Kraus JP, Doolittle RF, Konigsberg W, Rosenberg LE. Structure and expression of a complementary DNA for the nuclear coded precursor of human mitochondrial ornithine transcarbamylase. Science 1984; 224:1068–1074.
7. Davies KE, Briand P, Ionasescu V, Ionasescu G, Williamson R, Brown C, Cavard C, Cathelineau L. Gene for OTC: characterization and linkage to Duchenne muscular dystrophy. Nucl Acids Res 1985; 13:155–165.
8. Hata A, Tsuzuki T, Shimada K, Takiguchi M, Mori M, Matsuda I. Structure of the human ornithine transcarbamylase gene. J Biochem 1988; 103:302–308.

9. Tuchman M. The clinical, biochemical, and molecular spectrum of ornithine transcarbamylase deficiency. J Lab Clin Med 1992; 120:836–850.
10. Hata A, Tsuzuki T, Shimada K, Takiguchi M, Mori M, Matsuda I. Isolation and characterization of the human ornithine transcarbamylase gene: structure of the 5′-end region. J Biochem 1986; 100:717–725.
11. McIntyre P, Mercer JFB, Peterson G, Hudson P, Hoogenraad N. Selection of a cDNA clone which contains the complete coding sequence for the mature form of ornithine transcarbamylase from rat liver: expression of the cloned protein in *Escherichia coli*. Molecular cloning of rat ornithine transcarbamylase. Eur J Biochem 1984; 143:183–187.
12. Takiguchi M, Miura S, Mori M, Tatibana M, Nagata S, Kaziro Y. Molecular cloning and nucleotide sequence of cDNA for rat ornithine carbamoyltransferase precursor. Proc Natl Acad Sci USA 1984; 81:7412–7416.
13. Scherer SE, Veres G, Caskey CT. The genetic structure of mouse ornithine transcarbamylase. Nucl Acids Res 1988; 16:1593–1601.
14. Kraus JP, Hodges PE, Williamson CL, Horwich AL, Kalousek F, Williams KR, Rosenberg LE. A cDNA clone for the precursor of rat mitochondrial ornithine transcarbamylase: comparison of rat and human leader sequences and conservation of catalytic sites. Nucl Acids Res 1985; 13:943–952.
15. Veres G, Craigen WJ, Caskey CT. The 5′ flanking region of the ornithine transcarbamylase gene contains DNA sequences regulating tissue-specific expression. J Biol Chem 1986; 261:7588–7591.
16. Takiguchi M, Murakami T, Miura S, Mori M. Structure of the rat ornithine carbamoyltransferase gene, a large, X chromosome-linked gene with an atypical promoter. Proc Natl Acad Sci USA 1987; 84:6136–6140.
17. Ohtake A, Takiguchi M, Shigeto Y, Amaya Y, Kawamoto S, Mori M. Structural organization of the gene for rat liver-type arginase. J Biol Chem 1988; 263:2245–2249.
18. Engelhardt JF, Steel G, Valle D. Transcriptional analysis of the human ornithine aminotransferase promoter. J Biol Chem 1991; 266: 752–758.
19. Mori M, Murakami T, Haraguchi Y, Nishiyori A, Takiguchi M. Structure and expression of genes for urea cycle enzymes. In: Endou H, Schoolwerth AC, Baverel G, Tizianello A, editors. Molecular Aspects of Ammoniagenesis. Contrib Nephrol. Basel: Karger; 1991; 92: 218–223.
20. Kimura A, Nishiyori A, Murakami T, Tsukamoto T, Hata S, Osumi T, Okamura R, Mori M, Takiguchi M. Chicken ovalbumin upstream promoter-transcription factor (COUP-TF) represses transcription from the promoter of the gene for ornithine transcarbamylase in a manner antagonistic to hepatocyte nuclear factor-4 (HNF-4). J Biol Chem 1993; 268:11125–11133.
21. Murakami T, Nishiyori A, Takiguchi M, Mori M. Promoter and 11-kilobase upstream enhancer elements responsible for hepatoma cell-specific ex-

pression of the rat ornithine transcarbamylase gene. Molec Cell Biol 1990; 10:1180–1191.

22. Nishiyori A, Tashiro H, Kimura A, Akagi K, Yamamura K, Mori M, Takiguchi M. Determination of tissue specificity of the enhancer by combinatorial operation of tissue-enriched transcription factors. Both HNF-4 and C/EBPβ are required for liver-specific activity of the ornithine transcarbamylase enhancer. J Biol Chem 1994; 269: 1323–1331.
23. Huygen R, Crabeel M, Glansdorff N. Nucleotide sequence of the *ARG3* gene of the yeast *Saccharomyces cerevisiae* encoding ornithine carbamoyl transferase. Comparison with other carbamoyltransferases. Eur J Biochem 1987; 166:371–377.
24. DeRijke M, Seneca S, Punyammalee B, Glansdorff N, Crabeel M. Characterization of the DNA target site for the yeast *ARGR* regulatory complex, a sequence able to mediate repression or induction by arginine. Mol Cell Biol 1992; 12:68–81.
25. Boyce FM, Anderson GM, Rusk CD, Freytag SO. Human argininosuccinate synthetase minigenes are subject to arginine-mediated repression but not to *trans* induction. Mol Cell Biol 1986; 6:1244–1252.
26. Schimke RT. Studies on factors affecting the levels of urea cycle enzymes in rat liver. J Biol Chem 1963; 238:1012–1018.
27. Snodgrass PJ, Lin RC. Differing effects of arginine deficiency on the urea cycle enzymes of rat liver, cultured hepatocytes, and hepatoma cells. J Nutr 1987; 117: 1827–1837.
28. Schimke RT. Enzymes of arginine metabolism in mammalian cell culture. I. Repression of argininosuccinate synthetase and argininosuccinase. J Biol Chem 1964; 239:136–145.
29. Mullins LJ, Veres G, Caskey CT, Chapman V. Differential methylation of the ornithine carbamoyl transferase gene on active and inactive mouse X chromosomes. Mol Cell Biol 1987; 7:3916–3922.
30. Bencini DA, Houghton JE, Hoover TA, Folterman KF, Wild JR, O'Donovan GA. The DNA sequence of *arg I* from *Escherichia coli* K12. Nucl Acids Res 1983; 11:8509–8518.
31. Van Vliet F, Cunin R, Jacobs A, Piette J, Gigot D, Lauwereys M, Piérard A, Glansdorff N. Evolutionary divergence of genes for ornithine and aspartate carbamoyl-transferases—complete sequence and mode of regulation of the *Escherichia coli arg F* gene; comparison of *arg F* with *arg I* and *pyr B*. Nucl Acids Res 1984 12:6277–6288.
32. Upshall A, Gilbert T, Saari G, O'Hara PJ, Weglenski P, Berse B, Miller K, Timberlake WE. Molecular analysis of the *arg B* gene of *Aspergillus nidulans*. Mol Gen Genet 1986; 204:349–354.
33. Buxton FP, Gwynne DI, Garven S, Sibley S, Davies RW. Cloning and molecular analysis of the ornithine carbamoyl transferase gene of *Aspergillus niger*. Gene 1987; 60:255–266.

34. Baur H, Stalon V, Falmagne P, Luethi E, Haas D. Primary and quaternary structure of the catabolic ornithine carbamoyltransferase from *Pseudomonas aeruginosa.* Extensive sequence homology with the anabolic ornithine carbamoyltransferases of *Escherichia coli.* Eur J Biochem 1987; 166:111–117.
35. Itoh Y, Soldati L, Stalon V, Falmagne P, Terawaki Y, Leisinger T, Haas D. Anabolic ornithine carbamoyltransferase of *Pseudomonas aeruginosa:* nucleotide sequence and transcriptional control of the *arg F* structural gene. J Bact 1988; 170:2725–2734.
36. Martin PR, Cooperider JW, Mulks MH. Sequence of the *arg F* gene encoding ornithine transcarbamoylase from *Neisseria gonorrhoeae.* Gene 1990; 94:139–140.
37. Mountain A, Smith MCM, Baumberg S. Nucleotide sequence of the *Bacillus subtilis argF* gene encoding ornithine carbamoyltransferase. Nucl Acid Res 1990; 18:4594.
38. Skrzypek M, Borsuk P, Maleszka R. Cloning and sequencing of the ornithine carbamoyltransferase gene from *Pachysolen tannophilus.* Yeast 1990; 6:141–148.
39. Mosqueda G, Van den Broeck G, Saucedo O, Bailey AM, Alvarez-Morales A, Herrera-Estrella L. Isolation and characterization of the gene from *Pseudomonas syringae* pv. *phaseolicola* encoding the phaseolotoxin-insensitive ornithine carbamoyltransferase. Mol Gen Genet 1990; 222:461–466.
40. Hatziloukas E, Panopoulos NJ. Origin, structure, and regulation of *argK,* encoding the phaseolotoxin-resistant ornithine carbamoyltransferase in *Pseudomonas syringae* pv. phaseolicola, and functional expression of *argK* in transgenic tobacco. J Bact 1992; 174:5895–5909.
41. Timm J, Van Rompaey I, Tricot C, Massaer M, Haeseleer F, Fauconnier A, Stalon V, Bollen A, Jacobs P. Molecular cloning, characterization and purification of ornithine carbamoyltransferase from *Mycobacterium bovis* BCG. Mol Gen Genet 1992; 234:475–480.
42. Van Huffel C, Dubois E, Messenguy F. Cloning and sequencing of *arg 3* and *arg 11* genes of *Schizosaccharomyces pombe* on a 10-kb DNA fragment. Heterologous expression and mitochondrial targeting of their translation products. Eur J Biochem 1992; 205:33–43.
43. Ruepp A, Müller HN, Lottspeich F, Soppa J. Catabolic ornithine transcarbamylase of *Halobacterium halobium (salinarium):* purification, characterization, sequence determination, and evolution. J Bact 1995; 177:1129–1136.
44. Williamson CL, Lake MR, Slocum RD. Isolation and characterization of a cDNA encoding a pea ornithine transcarbamoylase *(argF)* and comparison with other transcarbamoylases. Plant Mol Biol 1996; 31:1087–1092.
45. Koger JB, Jones EE. Rapid communication: nucleotide sequence of porcine OTCase cDNA. J Anim Sci 1997; 75:3368.

46. Ventura L, Pérez-González JA, Ramón D. Cloning and molecular analysis of the *Aspergillus terreus argI* gene coding for an ornithine carbamoyltransferase. FEMS Microbiol Lett 1997; 149: 207–212.
47. Roovers M, Hethke C, Legrain C, Thomm M, Glansdorff N. Isolation of the gene encoding *Pyrococcus furiosus* ornithine carbamoyltransferase and study of its expression profile *in vivo* and *in vitro*. Eur J Biochem 1997; 247:1038–1045.
48. Sanchez R, Baetens M, Van de Casteele M, Roovers M, Legrain C, Glansdorff N. Ornithine carbamoyltransferase from the extreme thermophile *Thermus thermophilus*. Analysis of the gene and characterization of the protein. Eur J Biochem 1997; 248:466–474.
49. Shimogiri T, Kono M, Mannen H, Mizutani M, Tsuji S. Chicken ornithine transcarbamylase gene, structure, regulation, and chromosomal assignment: repetitive sequence motif in intron 3 regulates this enzyme activity. J Biochem 1998; 124:962–971.
50. Quesada V, Ponce MR, Micol JL. OTC and AUL1, two convergent and overlapping genes in the nuclear genome of *Arabidopsis thaliana*. FEBS Lett 1999; 461:101–106.
51. Chun J-Y, Lee M-S. Cloning of the *argF* gene encoding the ornithine carbamoyltransferase from *Corynebacterium glutamicum*. Mol Cells 1999; 9:333–337.

2. *Synthesis, Processing and Assembly*

52. Morita T, Mori M, Tatibana M, Cohen PP. Site of synthesis and intracellular transport of the precursor of mitochondrial ornithine carbamoyltransferase. Biochem Biophys Res Commun 1981; 99:623–629.
53. Conboy JG, Kalousek F, Rosenberg LE. *In vitro* synthesis of a putative precursor of mitochondrial ornithine transcarbamoylase. Proc Natl Acad Sci USA 1979; 76:5724–5727.
54. Kraus JP, Conboy JG, Rosenberg LE. Pre-ornithine transcarbamylase. Properties of the cytoplasmic precursor of a mitochondrial matrix enzyme. J Biol Chem 1981; 256:10739–10742.
55. Hachiya N, Komiya T, Alam R, Iwahashi J, Sakaguchi M, Omura T, Mihara K. MSF, a novel cytoplasmic chaperone which functions in precursor targeting to mitochondria. EMBO J 1994; 13:5146–5154.
56. Schatz G. The protein import system of mitochondria. J Biol Chem 1996; 271:31763–31766.
57. Terada K, Ohtsuka K, Imamoto N, Yoneda Y, Mori M. Role of heat shock cognate 70 protein in import of ornithine transcarbamylase precursor into mammalian mitochondria. Molec Cell Biol 1995; 15:3708–3713.
58. Kanazawa M, Terada K, Kato S, Mori M. HSDJ, a human homolog of DnaJ, is farnesylated and is involved in protein import into mitochondria. J Biochem 1997; 121:890–895.

59. Murakami K, Tanase S, Morino Y, Mori M. Presequence binding factor-dependent and -independent import of proteins into mitochondria. J Biol Chem 1992; 267: 13119–13122.
60. Schleyer M, Neupert W. Transport of proteins into mitochondria: translocational intermediates spanning contact sites between outer and inner membranes. Cell 1985; 43:339–350
61. González-Bosch C, Marcote MJ, Hernández-Yago J. Role of polyamines in the transport *in vitro* of the precursor of ornithine transcarbamylase. Biochem J 1991; 279:815–820.
62. Marcote MJ, Corella D, González-Bosch C, Hernández-Yago J. A structure-effect study of the induction by polyamines of the transport *in vitro* of the precursor of ornithine transcarbamylase. Biochem J 1994; 300:277–280.
63. Terada K, Kanazawa M, Yano M, Hanson B, Hoogenraad N, Mori M. Participation of the import receptor Tom20 in protein import into mammalian mitochondria: analyses in vitro and in cultured cells. FEBS Lett 1997; 403:309–312.
64. Nuttall SD, Hanson BJ, Mori M, Hoogenraad NJ. hTom34: a novel translocase for the import of proteins into human mitochondria. DNA Cell Biol 1997; 16:1067–1074.
65. Mori M, Miura S, Morita T, Takiguchi M, Tatibana M. Ornithine transcarbamylase in liver mitochondria. Molec Cell Biochem 1982; 49:97–111.
66. Miura S, Mori M, Tatibana M. Transport of ornithine carbamoyltransferase precursor into mitochondria. J Biol Chem 1983; 258:6671–6674.
67. Cheng MY, Hartl F-U, Martin J, Pollock RA, Kalousek F, Neupert W, Hallberg EM, Hallberg RL, Horwich AL. Mitochondrial heat-shock protein hsp60 is essential for assembly of proteins imported into yeast mitochondria. Nature 1989; 3377:620–625.
68. Hartman DJ, Hoogenraad NJ, Condron R, Høj PB. Identification of a mammalian 10-kDa heat shock protein, a mitochondrial chaperonin 10 homologue essential for assisted folding of trimeric ornithine transcarbamoylase *in vitro*. Proc Natl Acad Sci USA 1992; 89:3394–3398.
69. Mori M, Miura S, Tatibana M, Cohen PP. Characterization of a protease apparently involved in processing of pre-ornithine transcarbamylase of rat liver. Proc Natl Acad Sci USA 1980; 77:7044–7048.
70. Kalousek F, Hendrick JP, Rosenberg LE. Two mitochondrial matrix proteases act sequentially in the processing of mammalian matrix enzymes. Proc Natl Acad Sci USA 1988; 85:7536–7540.
71. Kolansky DM, Conboy JG, Fenton WA, Rosenberg LE. Energy-dependent translocation of the precursor of ornithine transcarbamylase by isolated rat liver mitochondria. J Biol Chem 1982; 257:8467–8471.
72. Conboy JG, Fenton WA, Rosenberg LE. Processing of pre-ornithine transcarbamylase requires a zinc-dependent protease localized to the mitochondrial matrix. Biochem Biophys Res Commun 1982; 105:1–7.

73. Isaya G, Kalousek F, Rosenberg LE. Amino-terminal octapeptides function as recognition signals for the mitochondrial intermediate peptidase. J Biol Chem 1992; 267:7904–7910.
74. Zheng X, Rosenberg LE, Kalousek F, Fenton WA. GroEL, GroES, and ATP-dependent folding and spontaneous assembly of ornithine transcarbamylase. J Biol Chem 1993; 268:7489–7493.
75. Horwich AL, Kalousek F, Rosenberg LE. Arginine in the leader peptide is required for both import and proteolytic cleavage of a mitochondrial precursor. Proc Natl Acad Sci USA 1985; 82:4930–4933.
76. Lingelbach KR, Graf LJ, Dunn AR, Hoogenraad NJ. Effect of deletions within the leader peptide of pre-ornithine transcarbamylase on mitochondrial import. Eur J Biochem 1986; 161:19–23
77. Graf L, Lingelbach K, Hoogenraad J, Hoogenraad N. Mitochondrial import of rat pre-ornithine transcarbamylase: accurate processing of the precursor form is not required for uptake into mitochondria, nor assembly into catalytically active enzyme. Prot Engineering 1988; 2:297–300.
78. Horwich AL, Kalousek F, Fenton WA, Pollock RA, Rosenberg LE. Targeting of pre-ornithine transcarbamylase to mitochondria: definition of critical regions and residues in the leader peptide. Cell 1986; 44:451–459.
79. Sztul ES, Hendrick JP, Kraus JP, Wall D, Kalousek F, Rosenberg LE. Import of rat ornithine transcarbamylase precursor into mitochondria: two step processing of the leader peptide. J Cell Biol 1987; 105:2631–2639.
80. Isaya G, Fenton WA, Hendrick JP, Furtak K, Kalousek F, Rosenberg LE. Mitochondrial import and processing of mutant human ornithine transcarbamylase precursors in cultured cells. Molec Cell Biol 1988; 8:5150–5158.
81. Epand RM, Hui S-W, Argan C, Gillespie LL, Shore GC. Structural analysis and amphiphilic properties of a chemically synthesized mitochondrial signal peptide. J Biol Chem 1986; 261:10017–10020.
82. Gillespie LL, Argan C, Taneja AT, Hodges RS, Freeman KB, Shore GC. A synthetic signal peptide blocks import of precursor proteins destined for the mitochondrial inner membrane or matrix. J Biol Chem 1985; 260:16045–16048.
83. Skerjanc IS, Sheffield WP, Silvius JR, Shore GC. Identification of hydrophobic residues in the signal sequence of mitochondrial preornithine carbamyltransferase that enhance the rate of precursor import. J Biol Chem 1988; 263:17233–17236.
84. Skerjanc IS, Sheffield WP, Randall SK, Silvius JR, Shore GC. Import of precursor proteins into mitochondria: site of polypeptide unfolding. J Biol Chem 1990; 265: 9444–9451.
85. Sheffield WP, Shore GC, Randall SK. Mitochondrial precursor protein. Effects of 70-kilodalton heat shock protein on polypeptide folding, aggregation, and import competence. J Biol Chem 1990; 265:11069–11076.
86. Côté C, Poirier J, Boulet D, Dionne G, Lacroix M. Structural identity be-

tween the NH2—terminal domain of the rat and human carbamyltransferase "targeting" sequences. J Biol Chem 1988; 263:5752–5756.

87. Yokota S, Mori M. Immunoelectron microscopical localization of ornithine transcarbamylase in hepatic parenchymal cells of the rat. Histochem J 1986; 18:451–457.
88. Powers-Lee SG, Mastico RA, Benayan M. The interaction of rat liver carbamoyl phosphate synthetase and ornithine transcarbamoylase with inner mitochondria membranes. J Biol Chem 1987; 262:15683–15688.

3. Molecular and Kinetic Characteristics

89. Snodgrass PJ. The effects of pH on the kinetics of human liver ornithine-carbamyl phosphate transferase. Biochemistry 1968; 7:3047–3051.
90. Pierson DL, Cox SL, Gilbert BE. Human ornithine transcarbamylase. Purification and characterization of the enzyme from normal liver and the liver of a Reye's syndrome patient. J. Biol Chem 1977; 252:6464–6469.
91. Kalousek F, Baudouin F, Rosenberg LE. Isolation and characterization of ornithine transcarbamylase from normal human liver. J Biol Chem 1978; 253:3939–3944.
92. Grisolia S, Cohen PP. The catalytic role of carbamyl glutamate in citrulline biosynthesis. J Biol Chem 1952; 198:561–572.
93. Reichard, P. Ornithine carbamyl transferase from rat liver. Acta Chem Scand 1957; 11:523–536.
94. Clarke S. The polypeptides of rat liver mitochondria: identification of a 36,000 dalton polypeptide as the subunit of ornithine transcarbamylase. Biochem Biophys Res Commun 1976; 71:1118–1124.
95. Lusty CJ, Jilka RL, Nietsch EH. Ornithine transcarbamylase of rat liver. Kinetic, physical, and chemical properties. J Biol Chem 1979; 254:10030–10036.
96. Hoogenraad NJ, Sutherland TM, Howlett GJ. Purification of ornithine transcarbamylase from rat liver by affinity chromatography with immobilized transition-state analog. Anal Biochem 1980; 101:97–102.
97. Burnett GH, Cohen PP. Study of carbamyl phosphate-ornithine transcarbamylase. J Biol Chem 1957; 229:337–344.
98. Joseph RL, Baldwin E, Watts DC. Studies on carbamoyl phosphate-L-ornithine carbamoyltransferase from ox liver. Biochem J 1963; 87:409–416.
99. Marshall M, Cohen PP. Ornithine transcarbamylase from *Streptococus faecalis* and bovine liver. I. Isolation and subunit structure. J Biol Chem 1972; 247:1641–1653.
100. Koger JB, Howell RG, Kelly M, Jones EE. Purification and properties of porcine liver ornithine transcarbamylase. Arch Biochem Biophys 1994; 309:293–299.
101. DeGregorio A, Valentini G, Bellocco E, Desideri A, Cuzzocrea G. A comparative study on liver ornithine carbamoyl transferase from a marine

mammal *Stenella* and an elasmobranch *Sphyrna zygaena*. Comp Biochem Physiol 1993; 105B: 497–501.

102. Brown GW Jr, Cohen PP. Comparative biochemistry of urea synthesis. 3. Activities of urea-cycle enzymes in various higher and lower vertebrates. Biochem J 1960; 75:82–91.
103. Mora J, Martuscelli J, Ortiz-Pineda J, Soberon G. The regulation of urea-biosynthesis enzymes in vertebrates. Biochem J 1965; 96:28–35.
104. Mommsen TP, Walsh PJ. Evolution of urea synthesis in vertebrates: the piscine connection. Science 1989; 243:72–75.
105. Baldwin E. Ureogenesis in elasmobranch fishes. Comp Biochem Physiol 1960; 1:24–37.
106. Goldstein L, Janssens PA, Forster RP. Lungfish *Neoceratodus forsteri:* activities of ornithine-urea cycle and enzymes. Science 1967; 157:316–317.
107. Chadwick TD, Wright PA. Nitrogen excretion and expression of urea cycle enzymes in the atlantic cod (*Gadus morhua* L.): a comparison of early life stages with adults. J Exp Biol 1999; 202:2653–2662.
108. Saha N, Ratha BK. Comparative study of ureogenesis in freshwater, air-breathing teleosts. J Exp Zool 1989; 252:1–8.
109. Randall DJ, Wood CM, Perry SF, Bergman H, Maloiy GMO, Mommsen TP, Wright PA. Urea excretion as a strategy for survival in a fish living in a very alkaline environment. Nature 1989; 337:165–166.
110. Tsuiji S. Chicken ornithine transcarbamylase: purification and some properties. J Biochem 1983; 94:1307–1315.
111. Ratner S. Enzymes of arginine and urea synthesis. Adv Enzymol 1973; 39:1–90.
112. De Ruiter H, Kollöffel C. Properties of ornithine carbamoyltransferase from *Pisum sativum* L. Plant Physiol 1985; 77:695–699.
113. Kleczkowski K, Cohen PP. Purification of ornithine transcarbamylase from pea seedlings. Arch Biochem Biophys 1964; 107:271–278.
114. Slocum RD, Richardson DP. Purification and characterization of ornithine transcarbamylase from pea (*Pisum sativum* L.) Plant Physiol 1991; 96:262–268.
115. Bellocco E, Leuzzi U, De Gregorio A, Laganà G, Risitano A, Desideri A, Cuzzocrea G. Partial purification and properties of ornithine carbamoyl-transferase from *Citrus limonum* leaves. Biochem Molec Biol Internat 1993; 29:281–290.
116. Lee Y, Jun BO, Kim S-G, Kwon YM. Purification of ornithine carbamoyl-transferase from kidney bean (*Phaseolus vulgaris* L.) leaves and comparison of the properties of the enzyme from canavanine-containing and -deficient plants. Planta 1998; 205:375–379.
117. Slocum RD, Nichols HFIII, Williamson CL. Purification and characterization of *Arabidopsis* ornithine transcarbamoylase (OTCase), a member of a distinct and evolutionarily-conserved group of plant OTCases. Plant Physiol Biochem 2000; 38:279–288.

118. Lee Y, Lee CB,, Kim S-G, Kwon YM. Purification and characterization of ornithine carbamoyltransferase from the chloroplasts of *Canavalia lineata* leaves. Plant Science 1997; 122:217–224.
119. Xiong X, Anderson PM. Purification and properties of ornithine carbamoyl transferase from liver of *Squalus acanthias*. Arch Biochem Biophys 1989; 270:198–207.
120. Eisenstein E, Osborne JCJr, Chaiken IM, Hensley P. Purification and characterization of ornithine transcarbamoylase from *Saccharomyces cerevisiae*. J Biol Chem 1984; 259:5139–5145.
121. Bates M, Weiss RL, Clarke S. Ornithine transcarbamylase from *Neurospora crassa:* purification and properties. Arch Biochem Biophys 1985; 239: 172–183.
122. De La Fuente JL, Martín JF,, Liras P. New type of hexameric ornithine carbamoyltransferase with arginase activity in the cephamycin producers *Streptomyces clavuligerus* and *Nocardia lactamdurans*. Biochem J 1996; 320:173–179.
123. Legrain C, Villleret V, Roovers M, Gigot D, Dideberg O, Piérard A, Glansdorff N. Biochemical characterization of ornithine carbamoyltransferase from *Pyrococcus furiosus*. Eur J Biochem 1997; 247:1046–1055.
124. Legrain C, Stalon V. Ornithine carbamoyltransferase from *Escherichia coli* W. Purification, structure and steady-state kinetic analysis. Eur J Biochem 1976; 63:289–301.
125. Stalon V, Ramos F, Piérard A, Wiame J-M. Regulation of the catabolic ornithine carbamoyltransferase of *Pseudomonas fluorescens*. A comparison with anabolic transferase and with a mutationally modified catabolic transferase. Eur J Biochem 1972; 29:25–35.
126. Legrain C, Stalon V, Noullez J-P, Mercenier A, Simon J-P, Broman K, Wiame J-M. Structure and function of ornithine carbamoyltransferases. Eur J Biochem 1977; 80:401–409.
127. Neway JO, Switzer RL. Purification, characterization, and physiological function of *Bacillus subtilis* ornithine transcarbamylase. J Bacteriol 1983; 155:512–521.
128. Abdelal ATH, Kennedy EH, Nainan O. Ornithine transcarbamylase from *Salmonella typhimurium:* purification, subunit composition, kinetic analysis, and immunological cross-reactivity. J Bact 1977; 129:1387–1396.
129. Ahmad S, Bhatnagar RK, Venkitasubramanian TA. Ornithine transcarbamylase from *Mycobacterium smegmatis* ATCC 14468: purification, properties, and reaction mechanism. Biochem Cell Biol 1986; 64:1349–1355.
130. Schimke RT, Berlin CM, Sweeney EW, Carroll WR. The generation of energy by the arginine dihydrolase pathway in *Mycoplasma hominis* 07. J Biol Chem 1966; 241:2228–2236.

131. Bishop SH, Grisolia S. Crystalline ornithine transcarbamylase. Biochim Biophys Acta 1967;139:344–348.
132. Tricot C, De Coen J-L, Momin P, Falmagne P, Stalon V. Evolutionary relationships among bacterial carbamoyltransferases. J Gen Microbiol 1989; 135:2453–2464.
133. Marshall M, Cohen PP. Ornithine transcarbamylase from *Streptococcus faecalis* and bovine liver. II. Multiple binding sites for carbamyl-P and L-norvaline, correlation with steady state kinetics. J Biol Chem 1972; 247:1654–1668.
134. Joseph RL, Watts DC, Baldwin E. The comparative biochemistry of carbamoyl phosphate-L-ornithine carbamoyltransferases in relation to their evolution. Comp Biochem Physiol 1964; 11:119–129.
135. Simon J-P, Stalon V. L-ornithine carbamoyltransferase from *Saccharomyces cerevisiae:* steady-state kinetic analysis. Eur J Biochem 1977; 75:571–581.
136. Kuo LC, Herzberg W, Lipscomb WN. Substrate specificity and protonation state of ornithine transcarbamoylase as determined by pH studies. Biochemistry 1985; 24:4754–4761.
137. Ravel JM, Grona ML, Humphreys JS, Shive W. Properties and biotin content of purified preparations of the ornithine-citrulline enzyme of *Streptococcus lactis.* J Biol Chem 1959; 234:1452–1455.
138. Sainz G, Tricot C, Foray M-F, Marion D, Dideberg O, Stalon V. Kinetic studies of allosteric catabolic ornithine carbamoyltransferase from *Pseudomonas aeruginosa.* Eur J Biochem 1998; 251:528–533.
139. Mouz N, Tricot C, Ebel C, Petillot Y, Stalon V, Dideberg O. Use of a designed fusion protein dissociates allosteric properties from the dodecameric state of *Pseudomonas aeruginosa* catabolic ornithine carbamoyltransferase. Proc Natl Acad Sci USA 1996; 93:9414–9419.
140. Tricot C, Nguyen VT, Stalon V. Steady-state kinetics and analysis of pH dependence on wild-type and a modified allosteric *Pseudomonas aeruginosa* ornithine carbamoyltransferase containing the replacement of glutamate 105 by alanine. Eur J Biochem 1993; 215:833–839.
141. Snodgrass PJ, Parry DJ. The kinetics of serum ornithine carbamoyltransferase. J Lab Clin Med 1969; 73:940–950.
142. Batchelder AC, Schmidt CLA. The effects of certain salts on the dissociation of aspartic acid, arginine, and ornithine. J Phys Chem 1940; 44:893–909.
143. Krebs HA, Eggleston LV, Knivett VA. Arsenolysis and phosphorolysis of citrulline in mammalian liver. Biochem J 1955; 59:185–193.
144. Costell M,Grisolia S. Decarbamoylating activity of ornithine transcarbamoylase. Biochem Biophys Res Commun 1985; 128: 441–448.
145. Wallace R, Knecht E, Grisolia S. Turnover of rat liver ornithine transcarbamylase. FEBS Lett 1986; 208: 427–430.
146. Mori M, Aoyagi K, Tatibana M, Ishikawa T, Ishii H. N^{δ}-(phosphonacetyl)-

L-ornithine, a potent transition state analogue inhihitor of ornithine carbamoyltransferase. Biochem Biophys Res Commun 1977; 76: 900–904.
147. Grillo MA. Inhibition of ornithine carbamoyltransferase by diethylpyrocarbonate. Enzymologia 1971; 40:265–271.
148. DeMars R, LeVan SL, Trend BL, Russell LB. Abnormal ornithine carbamoyltransferase in mice having the sparse-fur mutation. Proc Natl Acad Sci USA 1976; 73:1693–1697.
149. Veres G, Gibbs RA, Scherer SE, Caskey CT. The molecular basis of the sparse fur mouse mutation. Science 1987; 237:415–417.
150. Cooper A JL, Nieves E, Coleman AE, Filc-DeRicco S, Gelbard AS. Short-term fate of [^{13}N] ammonia in rat liver *in vivo*. J Biol Chem 1987; 262:1073–1080.
151. Cohen VS, Cheung C-W, Sijuwade E, Raijman L. Kinetic properties of carbamoyl-phosphate synthase (ammonia) and ornithine carbamoyltransferase in permeabilized mitochondria. Biochem J 1992; 282:173–180.
152. Matsuzawa T, Sugimoto N, Ishiguro I. Dynamism of rat liver ornithine metabolisms in relation to dietary high-protein stimulation of the urea cycle, Urea Cycle Diseases. Edited by A Lowenthal, A Mori, B Marescau. New York, Plenum Press, 1982, pp. 245–254.
153. Zollner H. Ornithine uptake by isolated hepatocytes and distribution within the cell. Int J Biochem 1984; 16:681–685.
154. Cohen NS, Cheung C-W, Raijman L. Channeling of extramitochondrial ornithine to matrix ornithine transcarbamylase. J Biol Chem 1987; 262:203–208.
155. Indiveri C, Tonazzi A, Stipani I, Palmieri F. The purified and reconstituted ornithine/citrulline carrier from rat liver mitochondria: electrical nature and coupling of the exchange reaction with H^+ translocation. Biochem J 1997; 327:349–356.
156. Kuo LC, Lipscomb WN, Kantrowitz ER. Zn(II)-induced cooperativity of *Escherichia coli* ornithine transcarbamoylase. Proc Natl Acad Sci USA 1982; 79:2250–2254.
157. Kuo LC, Caron C, Lee S, Herzberg W. Zn^{2+} regulation of ornithine transcarbamoylase. II. Metal binding site. J Mol Biol 1990; 211: 271–280.
158. Lee S, Shen W-H, Miller AW, Kuo LC. Zn^{2+} regulation of ornithine transcarbamoylase. I. Mechanisms of action. J Mol Biol 1990; 211:255–269.

4. *Active Site and Other Essential Residues*

159. Monaco HL, Crawford JL, Lipscomb WN. Three-dimensional structures of aspartate carbamoyltransferase from *Escherichia coli* and of its complex with cytidine triphosptate. Proc Natl Acad Sci USA 1978; 75: 5296–5280.
160. Honzatko RB, Lipscomb WN. Interactions of phosphate ligands with *Escherichia coli* aspartate carbamoyltransferase in the crystalline state. J Mol Biol 1982; 160: 265–286.

161. Krause KL, Volz KW, Lipscomb WN. 2.5Å structure of aspartate carbamoyltransferase complexed with the bisubstrate analog N-(phosphonacetyl)-L-aspartate.J Mol Biol 1987; 193: 527–553.
162. Ke H, Lipscomb WN, Cho Y, Honzatko RB. Complex of N-phosphonacetyl-L-aspartate with aspartate carbamoyltransferase. X-ray refinement, analysis of conformational changes and catalytic and allosteric mechanisms. J Mol Biol 1988; 204: 725–747.
163. Gouaux JE, Lipscomb WN. Three-dimensional structure of carbamoyl phosphate and succinate bound to aspartate carbamoyltransferase. Proc Natl Acad Sci USA 1988; 85: 4205–4208.
164. Gouaux JE, Lipscomb WN. Crystal structures of phosphonoacetamide ligated T and phosphonoacetamide and malonate ligated R states of aspartate carbamoyltransferase at 2.8-A resolution and neutral pH. Biochemistry 1990; 29: 389–402.
165. Jin L, Stec B, Lipscomb WN, Kantrowitz ER. Insights into the mechanisms of catalysis and heterotropic regulation of *Escherichia coli* aspartate transcarbamoylase based upon a structure of the enzyme complexed with the bisubstrate analogue N-phosphonacetyl-L-aspartate at 2.1 Å. Proteins:Struct Funct Genet 1999; 37: 729–742.
166. Endrizzi JA, Beernink PT, Alber T, Schachman HK. Binding of bisubstrate analog promotes large structural changes in the unregulated catalytic trimer of aspartate transcarbamoylase: implications for allosteric regulation. Proc Natl Acad Sci USA 2000; 97: 5077–5082.
167. Villeret V, Tricot C, Stalon V, Dideberg O. Crystal structure of *Pseudomonas aeruginosa* catabolic ornithine transcarbamoylase at 3.0-Å resolution: A different oligomeric organization in the transcarbamoylase family. Proc Natl Acad Sci USA 1995; 92: 10762–10766
168. Jin L, Seaton BA, Head JF. Crystal structure at 2.8 Å resolution of anabolic ornithine transcarbamylase from *Escherichia coli.* Nature Struct Biol 1997; 4: 622–625.
169. Ha Y, McCann MT, Tuchman M, Allewell NM. Substrate-induced conformational change in a trimeric ornithine transcarbamylase. Proc Natl Acad Sci USA 1997; 94: 9550–9555.
170. Tuchman M, Morizono H, Rajagopal BS, Plante RJ, Allewell NM. The biochemical and molecular spectrum of ornithine transcarbamylase deficiency. J Inher Metab Dis 1998; 21 (Suppl 1): 40–58.
171. Shi D, Morizono H, Ha Y, Aoyagi M, Tuchman M, Allewell NM. 1.85-Å resolution crystal structure of human ornithine transcarbamoylase complexed with N-phosphonacetyl-L-ornithine. Catalytic mechanism and correlation with inherited deficiency. J Biol Chem 1998; 273: 34247–34254.
172. Labedan B, Boyen A, Baetens M, Charlier D, Chen P, Cunin R, Durbeco V, Glansdorff N, Herve G, Legrain C, Liang Z, Purcarea C, Roovers M, Sanchez R, Toong T-L, Van de Casteele M, van Vliet F, Zhang Y-F. The evo-

lutionary history of carbamoyltransferases: a complex set of paralogous genes was already present in the last universal common ancestor. J Mol Evol 1999; 49: 461–473.
173. Ibid. (Supplementary material to ref. 172) Complete multiple alignment of the 64 sequences of carbamoyltransferases.
174. Sali A, Blundell TL. Comparative protein modelling by satisfaction of spatial restraints (Modeller Release 4). J Mol Biol 1993; 234: 779–815.
175. Murata LB, Schachman HK. Structural similarity between ornithine and aspartate transcarbamoylases of *Escherichea coli:* characterization of the active site and evidence for an inter-domain carboxy-terminal helix in ornithine transcarbamoylase. Prot Sci 1996; 5: 709–718.
176. Houghton JE, Bencini DA, O'Donovan GA, Wild JR. Protein differentiation: a comparison of aspartate transcarbamoylase and ornithine transcarbamoylase from *Escherichia coli* K-12. Proc Natl Acad Sci USA 1984; 81: 4864–4868.
177. Shi D, Morizono H, Aoyagi M, Tuchman M, Allewell NM. Crystal structure of human ornithine transcarbamylase complexed with carbamoyl phosphate and L-norvaline at 1.9 Å resolution. Proteins:Struct Funct Genet 2000; 39: 271–277.
178. Xi XG, Van Vliet F, Ladjimi MM, Cunin R, Herve,G. The catalytic site of *Escherichia coli* aspartate transcarbaamylase: interaction between histidine 134 and the carbonyl group of the substrate carbamyl phosphate. Biochemistry 1990; 29: 8491–8498.
179. Kleanthous C, Wemmer DE, Schachman HK. The role of an active site histidine in the catalytic mechanism of aspartate transcarbamoylase. J Biol Chem 1988; 263: 13062–13067.
180. Waldrop GL, Turnbull JL, Parmentier LE, O'Leary MH, Cleland WN, Schachman HK. Steady-state kinetics and isotope effects on the mutant catalytic trimer of aspartate transcarbamoylase containing the replacement of histidine 134 by alanine. Biochemistry 1992; 31: 6585–6591.
181. Lauritzen AM, Landfear SM, Lipscomb WB. Inactivation of the catalytic subunit of aspartate transcarbamylase by nitration with tetranitromethane. J Biol Chem 1980; 255: 602–607.
182. Xu W, Kantrowitz ER. Function of threonine-55 in the carbamoyl phosphate binding site of *Escherichia coli* aspartate transcarbamoylase. Biochemistry 1989; 28: 9937–9943.
183. Stebbins JW, Xu W, Kantrowitz ER. Three residues involved in binding and catalysis in the carbamyl phosphate binding site of *Escherichia coli* aspartate transcarbamylase. Biochemistry 1989; 28: 2592–2600.
184. Xu W, Kantrowitz ER. Function of serine-52 and serine-80 in the catalytic mechanism of *Escherichia coli* aspartate transcarbamoylase. Biochemistry 1991; 30: 2535–2542.
185. Newton CJ, Stevens RC, Kantrowitz ER. Importance of a conserved residue, aspartate-162, for the function of *Escherichia coli* aspartate transcarbamoylase. Biochemistry 1992; 31: 3026–3032.

186. Middleton SA, Stebbins JW, Kantrowitz ER. A loop involving catalytic chain residues 230–245 is essential for the stabilization of both allosteric forms of *Escherichia coli* aspartate transcarbamoylase. Biochemistry 1989; 28: 1617–1626.
187. Baker DP, Kantrowitz ER. The conserved residues glutamate-37, aspartate-100, and arginine-269 are important for the structural stabilization of *Escherichia coli* aspartate transcarbamoylase. Biochemistry 1993; 32: 10150–10158.
188. Baker DP, Stebbins JW, DeSena E, Kantrowitz ER. Glutamic acid 86 is important for positioning the 80's loop and arginine 54 at the active site of *Escherichia coli* aspartate transcarbamoylase and for the structural stabilization of the C1-C2 interface. J Biol Chem 1994; 269: 24608–24614.
189. Newton CJ, Kantrowitz ER. Importance of domain closure for homotropic cooperativity in *Escherichia coli* aspartate transcarbamylase. Biochemistry 1990; 29: 1444–1451.
190. Ha Y, Allewell NM. Intersubunit hydrogen bond acts as a global molecular switch in *Escherichia coli* aspartate transcarbamoylase. Proteins: Struct, Funct, Genet 1998; 33: 430–443.
191. Marshall M, Cohen PP. Ornithine transcarbamoylase from *Streptococcus faecalis* and bovine liver. III.Effects of chemical modifications of specific residues on ligand binding and enzymatic activity. J Biol Chem 1972; 247: 1669–1682.
192. Marshall M. Ornithine transcarbamylase from bovine liver, The Urea Cycle. Edited by S. Grisolia, R. Baguena, F. Mayor. New York, John Wiley and Sons, 1976, pp. 169–179.
193. Marshall M, Cohen PP. Ornithine transcarbamylases. Ordering of S-cyano peptides and location of characteristically reactive cysteinyl residues within the sequence. J Biol Chem 1980; 255: 7287–7290.
194. Marshall M, Cohen PP. The essential sulfhydryl group of ornithine transcarbamylases. Reaction with anionic, aromatic disulfides and properties of its cyano derivative. J Biol Chem 1980; 255: 7291–7295.
195. Marshall M, Cohen PP. The essential sulfhydryl group of ornithine transcarbamylases. pH dependence of the spectra of its 2-mercuri-4-nitrophenol derivative. J Biol Chem 1980; 255: 7296–7300.
196. Goldsmith JO, Lee S, Zambidis I, Kuo LC. Control of L-ornithine specificity in *Escherichia coli* ornithine transcarbamoylase. Site-directed mutagenic and pH studies. J Biol Chem 1991; 266: 18626–18634.
197. McDowall S, Van Heeswijck R, Hoogenraad N. Site-directed mutagenesis of Arg 60 and Cys 271 in ornithine transcarbamylase from rat liver. Prot Engineer 1990; 4: 73–77.
198. Grillo MA, Bedino S. Ornithine carbamyltransferase of bovine liver. Enzymologia 1968; 35: 1–10.
199. Grillo MA, Coghe M. Tyrosine at the active centre of ornithine carbamyltransferase. Ital J Biochem 1969; 18:133–137.
200. Marshall M, Cohen PP. Kinetics and equilibrium of the inactivation of or-

nithine transcarbamylases by pyridoxal 5′-phosphate. J Biol Chem 1977; 252: 4276–4286.

201. Kalousek F, Rosenberg LE. Essential arginine and lysine residues in ornithine transcarbamylase from human and bovine liver. Fed Proc 1978; 37: 1310A.

202. Valentini G, DeGregorio A, DiSalvo C, Grimm R, Bellocco E, Cuzzocrea G, Iadarola P. An essential lysine in the substrate-binding site of ornithine carbamoyltransferase. Eur J Biochem 1996; 239: 397–402.

203. Marshall M, Cohen PP. Evidence for an exceptionally reactive arginyl residue at the binding site for carbamyl phosphate in bovine ornithine transcarbamylase. J Biol Chem 1980; 255: 7301–7305.

204. Fortin AF, Hauber JM, Kantrowitz ER. Comparison of the essential arginine residue in *Escherichia coli* ornithine and aspartate transcarbamylases. Biochim Biophys Acta 1981; 662: 8–14.

205. Kuo LC, Miller AW, Lee S, Kozuma C. Site-directed mutagenesis of *Escherichia coli* ornithine transcarbamoylase: role of arginine-57 in substrate binding and catalysis. Biochemistry 1988; 27: 8823–8832.

206. Goldsmith JO, Kuo LC. Protonation of arginine 57 of *Escherichia coli* ornithine transcarbamoylase regulates substrate binding and turnover. J Biol Chem 1993; 268: 18485–18490.

207. Rynkiewicz MJ, Seaton BA. Chemical rescue by guanidine derivatives of an arginine-substituted site-directed mutant of *Escherichia coli* ornithine transcarbamylase. Biochemistry 1996; 35: 16174–16179.

208. Zambidis I, Kuo LC. Substrate specificity and protonation state of *Escherichia coli* ornithine transcarbamoylase as determined by pH studies.Binding of carbamoyl phosphate. J Biol Chem 1990; 265: 2620–2623.

209. Kuo LC, Zambidis I, Caron C. Triggering of allostery in an enzyme by a point mutation: ornithine transcarbamoylase. Science 1989; 245: 522–524.

210. Shen WH. Fluorescence lifetimes of the tryptophan residues in ornithine transcarbamoylase. Biochemistry 1993; 32: 13925–13932.

211. Gouaux JE, Krause KL, Lipscomb WN. The catalytic mechanism of *Escherichia coli* aspartate carbamoyltransferase: a molecular modelling study. Biochem Biophys Res Commun 1987; 142: 893–897.

212. Stark GR. Aspartate transcarbamylase. The use of primary kinetic and solvent deuterium isotope effects to delineate some aspects of the mechanism. J Biol Chem 1971; 246: 3064–3068.

213. Waldrop GL, Urbauer JL, Cleland WW. ^{15}N isotope effects on nonenzymatic and aspartate transcarbamylase catalyzed reactions of carbamyl phosphate. J Am Chem Soc 1992; 114: 5941–5945.

214. Parmentier LE, O'Leary MH, Schachman HK, Cleland WW. ^{13}C isotope effects as a probe of the kinetic mechanism and allosteric properties of *Escherichia coli* aspartate transcarbamylase. Biochemistry 1992; 31: 6570–6576.

215. Parmentier LE, Weiss PM, O'Leary MH, Schachman HK, Cleland WW.

^{13}C and ^{15}N isotope effects as a probe of the chemical mechanism of *Escherichia coli* aspartate transcarbamylase. Biochemistry 1992; 31: 6577–6584.

216. Miller AW, Kuo LC. Ligand-induced isomerizations of *Escherichia coli* ornithine transcarbamoylase. An ultraviolet difference analysis. J Biol Chem 1990; 265: 15023–15027.
217. Kuo LC, Seaton BA. X-ray diffraction analysis on single crystals of recombinant *Escherichia coli* ounithine transcarbamoylase. J Biol Chem 1989; 264: 16246–16248.
218. Parmentier LE, Smith Kristensen J. Studies on the urea cycle enzyme ornithine transcarbamylase using heavy atom isotope effects. Biochim Biophys Acta 1998; 1382: 333–338.

5. Molecular Pathology of OTC Deficiency

219. Rodeck CH, Patrick AD, Pembrey ME, Tzanotos C, Whitfield AE. Fetal liver biopsy for prenatal diagnosis of ornithine carbamyltransferase deficiency. Lancet 1982; 2:297–300.
220. Holzgreve W, Golbus MS. Prenatal diagnosis of ornithine transcarbamylase deficiency utilizing fetal liver biopsy. Am J Hum Genet 1984; 36:320–328.
221. Shulman LP, Elias S. Percutaneous umbilical blood sampling, fetal skin sampling, and fetal liver biopsy. Semin Perinatol 1990; 14:456–464.
222. Francke U. Random X inactivation resulting in mosaic nullisomy of region Xp21.1→p21.3 associated with heterozygosity for ornithine transcarbamylase deficiency and for chronic granulomatous disease. Cytogenet Cell Genet 1984; 38:298–307.
223. Maddalena A, Sosnoski DM, Berry GT, Nussbaum RL. Mosaicism for an intragenic deletion in a boy with mild ornithine transcarbamylase deficiency. New Engl J Med 1988; 319:999–1003.
224. Slomski R, Braulke I, Behrend C, Schroder E, Colombo J-P, Reiss J. Ornithine transcarbamylase (OTC) deficiency in a female patient with a de novo deletion of the paternal X chromosome. Hum Genet 1992; 89:632–634.
225. Old JM, Purvis-Smith S, Wilcken B, Pearson P, Williamson R, Briand PL, Howard NJ, Hammond J, Cathelineau L, Davies KE. Prenatal exclusion of ornithine transcarbamylase deficiency by direct gene analysis. Lancet 1985; I:73–75.
226. Rozen R, Fox J, Fenton WA, Horwich AL, Rosenberg LE. Gene deletion and restriction fragment length polymorphisms at the human ornithine transcarbamylase locus. Nature 1985; 313:815–817.
227. Pembrey ME, Old JM, Leonard JV, Rodeck CH, Warren R, Davies KE. Prenatal diagnosis of ornithine carbamoyl transferase deficiency using a gene specific probe. J Med Genet 1985; 22:462–465.
228. McClead RE Jr, Rozen R, Fox J, Rosenberg L, Menke J, Bickers R, Mor-

row G III. Clinical application of DNA analysis in a family with OTC deficiency. Am J Med Genet 1986; 25:513–518.
229. Fox J, Hack AM, Fenton WA, Golbus MS, Winter S, Kalousek F, Rosen R, Brusilow SW, Rosenberg LE. Prenatal diagnosis of ornithine transcarbamylase deficiency with use of DNA polymorphisms. New Engl J Med 1986; 315:1205–1208.
230. Nussbaum RL, Boggs BA, Beaudet AL, Doyle S, Potter JL, O'Brien WE. New mutation and prenatal diagnosis in ornithine transcarbamylase deficiency. Am J Hum Genet 1986; 38:149–158.
231. Fox JE, Hack AM, Fenton WA, Rosenberg LE. Identification and application of additional restriction fragment length polymorphisms at the human ornithine transcarbamylase locus. Am J Hum Genet 1986; 38:841–847.
232. Spence JE, Maddalena A, O'Brien WE, Fernbach SD, Batshaw ML, Leonard CO, Beaudet AL. Prenatal diagnosis and heterozygote detection by DNA analysis in ornithine transcarbamylase deficiency. J Pediatr 1989; 114:582–588.
233. Matsuda I, Hata A, Matsuura T, Tsuzuki T, Shimada K. Structure of the ornithine transcarbamylase (OTC) gene and DNA diagnosis of OTC deficiency. Clin Chim Acta 1989; 185:283–290.
234. Liechti-Gallati S, Dionisi C, Bachmann C, Wermuth B, Colombo JP. Direct and indirect mutation analyses in patients with ornithine transcarbamylase deficiency. Enzyme 1991; 45:81–91.
235. Bale AE. Prenatal diagnosis of ornithine transcarbamylase deficiency. Prenat Diagn 1999; 19:1052–1054.
236. Watanabe A, Sekizawa A, Taguchi A, Saito H, Yanaihara T, Shimazu M, Matsuda I. Prenatal diagnosis of ornithine transcarbamylase deficiency by using a single nucleated erythrocyte from maternal blood. Hum Genet 1998; 102:611–615.
237. Hentzen D, Pelet A, Feldman D, Rabier D, Berthelot J, Munnich A. Fatal hyperammonemia resulting from a C-to-T mutation at a MspI site of the ornithine transcarbamylase gene. Hum Genet 1991; 88:153–156.
238. Bonaiti-Pellié C, Pelet A, Ogier H, Nelson J-R, Largillière C, Berthelot J, Saudubray J-M, Munnich A. A probable sex difference in mutation rates in ornithine transcarbamylase deficiency. Hum Genet 1990; 84:163–166.
239. Tuchman M, Matsuda I, Munnich A, Malcolm S, Strautnieks S, Briede T. Proportions of spontaneous mutations in males and females with ornithine transcarbamylase deficiency. Amer J Med Genet 1995; 55:67–70.
240. Grompe M, Caskey CT, Fenwick RG. Improved molecular diagnostics for ornithine transcarbamylase deficiency. Am J Hum Genet 1991; 48:212–222.
241. Hauser ER, Finkelstein JE, Valle D, Brusilow SW. Allopurinol induced orotidinuria: a test for mutations at the ornithine transcarbamylase locus in women. New Engl J Med 1990; 322:1641–1645.
242. Finkelstein JE, Francomano CA, Brusilow SW, Traystman MD. Use of denaturing gradient gel electrophoresis for detection of mutation and pro-

spective diagnosis in late onset ornithine transcarbamylase deficiency. Genomics 1990; 7:167–173.

243. Orita M, Iwahana H, Kanazawa H, Hayashi K, Sekiya T. Detection of polymorphisms of human DNA by gel electrophoresis as single-strand conformation polymorphisms. Proc Natl Acad Sci USA 1989; 86:2766–2770.
244. Grompe M, Muzny DM, Caskey CT. Scanning detection of mutations in human ornithine transcarbamoylase by chemical mismatch cleavage. Proc Natl Acad Sci USA 1989; 86:5888–5892.
245. Lee JT, Nussbaum RL. An arginine to glutamine mutation in residue 109 of human ornithine transcarbamylase completely abolishes enzymatic activity in Cos 1 cells. J Clin Invest 1989; 84:1762–1766.
246. Maddalena A, Spence JE, O'Brien WE, Nussbaum RL. Characterization of point mutations in the same arginine codon in three unrelated patients with ornithine transcarbamylase deficiency. J Clin Invest 1988; 82:1353–1358.
247. Hata A, Setoyama C, Shimada K, Takeda E, Kuroda Y, Akaboshi I, Matsuda I. Ornithine transcarbamylase deficiency resulting from a C-to-T substitution in exon 5 of the ornithine transcarbamylase gene. Am J Hum Genet 1989; 45:123–127.
248. Hata A, Matsuura T, Setoyama C, Shimada K, Yokoi T, Akaboshi I, Matsuda I. A novel missense mutation in exon 8 of the ornithine transcarbamylase gene in two unrelated male patients with mild ornithine transcarbamylase deficiency. Hum Genet 1991; 87:28–32.
249. Strautnieks S, Rutland P, Malcolm S. Arginine 109 to glutamine mutation in a girl with ornithine carbamoyl transferase deficiency. J Med Genet 1991; 28:871–874.
250. Carstens RP, Fenton WA, Rosenberg LE. Identification of RNA splicing errors resulting in human ornithine transcarbamylase deficiency. Am J Hum Genet 1991; 48:1105–1114.
251. Feldmann D, Rozet J-M, Pelet A, Hentzen D, Briand P, Hubert P, Largilliere C, Rabier D, Farriaux J-P, Munnich A. Site specific screening for point mutations in ornithine transcarbamylase deficiency. J Med Genet 1992; 29:471–475.
252. Suess PJ, Tsai MY, Holzknecht RA, Horowitz M, Tuchman M. Screening for gene deletions and known mutations in 13 patients with ornithine transcarbamylase deficiency. Biochem Med Metab Biol 1992; 47:250–259.
253. Satoh Y, Sannomyia Y, Ohtake A, Takayanagi M, Niimi H. Molecular characterization of four different mutant alleles of the OTC gene in four unrelated patients with ornithine transcarbamylase deficiency. J Jpn Pediatr Soc 1992; 96:1855–1862.
254. Tuchman M, Holzknecht RA, Gueron AB, Berry SA, Tsai MY. Six new mutations in the ornithine transcarbamylase gene detected by single-strand conformational polymorphism. Pediatr Res 1992; 32:600–604.
255. Matsuura T, Hoshide R, Setoyama C, Shimada K, Hase Y, Yanagawa T,

Kajita M, Matsuda I. Four novel gene mutations in five Japanese male patients with neonatal or late onset OTC deficiency: application of PCR-single-strand conformation polymorphisms for all exons and adjacent introns. Hum Genet 1993; 92:49–56.

256. Matsuura T, Hoshide R, Fukushima M, Sakiyama T, Owada M, Matsuda I. Prenatal monitoring of ornithine transcarbamoylase deficiency in two families by DNA analysis. J Inher Metab Dis 1993; 16:34–38.
257. Tuchman M. Mutations and polymorphisms in the human ornithine transcarbamylase gene. Hum Mutation 1993; 2:174–178.
258. Hoshide R, Matsuura T, Komaki S, Koike E, Ueno I, Matsuda I. Specificity of PCR-SSCP for detection of the mutant ornithine transcarbamylaese (OTC) gene in patients with OTC deficiency. J Inher Metab Dis 1993; 16:857–862.
259. Strautnieks S, Malcolm S. Novel mutation affecting a splice site in exon 4 of the ornithine carbamoyl transferase gene. Hum Molec Genet 1993: 2:1963–1964.
260. Reish O, Plante RJ, Tuchman M. Four new mutations in the ornithine transcarbamylase gene. Biochem Med Metab Biol 1993; 50:169–175.
261. Tsai MY, Holzknecht RA, Tuchman M. Single-strand conformational polymorphism and direct sequencing applied to carrier testing in families with ornithine transcarbamylase deficiency. Hum Genet 1993; 91:321–325.
262. Matsuura T, Hoshide R, Kiwaki K, Komaki S, Koike E, Endo F, Oyanagi K, Suzuki Y, Kato I, Ishikawa K, Yoda H, Kamitani S, Sakaki Y, Matsuda I. Four newly identified ornithine transcarbamylase (OTC) mutations (D126G, R129H, I172M and W332X) in Japanese male patients with early-onset OTC deficiency. Hum Mutation 1994; 3:402–406.
263. Tuchman M, Plante RJ, Giguere Y, Lemieux B. The ornithine transcarbamylase gene: new "private" mutations in four patients and study of a polymorphism. Hum Mutation 1994; 3:318–320.
264. Tuchman M, Plante RJ, McCann MT, Qureshi AA. Seven new mutations in the human ornithine transcarbamylase gene. Hum Mutation 1994; 4:57–60.
265. Matsuura T, Hoshide R, Setoyama C, Komaki S, Kiwaki K, Endo F, Nishikawa S, Matsuda I. Expression of four mutant human ornithine transcarbamylase genes in cultured Cos 1 cells relates to clinical phenotypes. Hum Genet 1994; 93:129–134.
266. Gilbert-Dussardier B, Rabier D, Strautnieks S, Segues B, Bonnefont JP, Munnich A. A novel arginine (245) to glutamine change in exon 8 of the ornithine carbamoyl transferase gene in two unrelated children presenting with late onset deficiency and showing the same enzymatic pattern. Hum Molec Genet 1994; 3:831–832.
267. Oppliger Liebundgut E, Liechti-Gallati S, Colombo J-P, Wermuth B. Ornithine transcarbamylase deficiency: new sites with increased probability of mutation. Hum Genet 1995; 95:191–196.

268. Tuchman M, Morizono H, Reish O, Yuan X, Allewell NM. The molecular basis of ornithine transcarbamylase deficiency: modelling the human enzyme and the effects of mutations. J Med Genet 1995; 32:680–688.
269. Zimmer K-P, Matsuura T, Colombo J-P, Koch H-G, Ullrich K, Deufel T, Harms E, Matsuda I. A novel point mutation at codon 269 of the ornithine transcarbamylase (OTC) gene causing neonatal onset of OTC deficiency. J Inher Metab Dis 1995; 18:356–357.
270. Tuchman M, Plante RJ. Mutations and polymorphisms in the human ornithine transcarbamylase gene: mutation update addendum. Hum Mutation 1995; 5:293–295.
271. García-Pérez MA, Sanjurjo P, Rubio V. Demonstration of the spf-ash mutation in Spanish patients with ornithine transcarbamylase deficiency of moderate severity. Hum Genet 1995; 95:183–186.
272. García-Pérez MA, Sanjurjo P, Briones P, Garcia-Muñoz MJ, Rubio V. A splicing mutation, a nonsense mutation (Y167X) and two missense mutations (I159T and A209V) in Spanish patients with ornithine transcarbamylase deficency. Hum Genet 1995; 96:549–551.
273. Matsuura T, Hoshide R, Komaki S, Kiwaki K, Endo F, Nakamura S, Jitosho T, Matsuda I. Identification of two new aberrant splicings in the ornithine carbamoyltransferase (OCT) gene in two patients with early and late onset OCT deficiency. J Inher Metab Dis 1995; 18:273–282.
274. Oppliger Leibundgut E, Wermuth B, Colombo J-P, Liechti-Gallati S. Identification of four novel splice site mutations in the ornithine transcarbamylase gene. Hum Genet 1996; 97:209–213.
275. Vella S, Steiner F, Schlumbom V, Zurbrügg R, Wiesmann UN, Schaffner T, Wermuth B. Mutation of ornithine transcarbamylase (H136R) in a girl with severe intermittent orotic aciduria but normal enzyme activity. J Inher Metab Dis 1996; 19:517–523.
276. Tuchman M, Plante RJ, Garcia-Perez MA, Rubio V. Relative frequency of mutations causing ornithine transcarbamylase deficiency in 78 families. Hum Genet 1996; 97:274–276.
277. Hoshide R, Matsuura T, Sagara Y, Kubo T, Shimadzu M, Endo F, Matsuda I. Prenatal monitoring in a family at high risk for ornithine transcarbamylase (OTC) deficiency: a new mutation of an A-to-C transversion in position +4 of intron 1 of the OTC gene that is likely to abolish enzyme activity.Amer J Med Genet 1996; 64:459–464.
278. Schimanski U, Krieger D, Horn M, Stremmel W, Wermuth B, Theilmann L. A novel two-nucleotide deletion in the ornithine transcarbamylase gene causing fatal hyperammonia in early pregnancy. Hepatol 1996; 24:1413–1415.
279. Yoo H-W, Kim G-H, Lee D-H. Identification of new mutations in the ornithine transcarbamylase (OTC) gene in Korean families. J Inher Metab Dis 1996; 19:31–42.
280. Matsuda I, Matsuura T, Nishiyori A, Komaki S, Hoshide R, Matsumoto T, Funakoshi M, Kiwaki K, Endo F, Hata A, Shimadzu M, Yoshino M.

Phenotypic variability in male patients carrying the mutant ornithine transcarbamylase (OTC) allele, Arg40His, ranging from a child with an unfavourable prognosis to an asymptomatic older adult. J Med Genet 1996; 33:645–648.

281. Gilbert-Dussardier B, Segues B, Rozet J-M, Rabier D, Calvas P, de Lumley L, Bonnefond J-P, Munnich A. Partial duplication [dup. TCAC(178)] and novel point mutations (T125M, G188R, A209V, and H302L) of the ornithine transcarbamylase gene in congenital hyperammonemia. Hum Mutation 1996; 8:74–76.
282. Oppliger Leibundgut E, Wermuth B, Colombo J-P, Liechti-Gallati S. Ornithine transcarbamylase deficiency: characterization of gene mutations and polymorphisms. Hum Mutation 1996; 8:333–339.
283. Ségues B, Veber PS, Rabier D, Calvas P, Saudubray J-M, Gilbert-Dussardier B, Bonnefont J-P, Munnich A. A 3-base pair in-frame deletion in exon 8 (delGlu272/273) of the ornithine transcarbamylase gene in late-onset hyperammonemic coma. Hum Mutation 1996; 8:373–374.
284. van Diggelen OP, Zaremba J, He W, Keulemans JLM, Boer AM, Reuser AJJ, Ausems MGEM, Smeitink JAM, Kowalczyk J, Pronicka E, Rokicki D, Tarnowska-Dziduszko E, Knepppers ALJ, Bakker E. Asymptomatic and late-onset ornithine transcarbamylase (OTC) deficiency in males of a five-generation family, caused by an A208T mutation. Clin Genet 1996; 50:310–316.
285. Tuchman M, Morizono H, Rajagopal BS, Plante RJ, Allewell NM. Identification of 'private' mutations in patients with ornithine transcarbamylase deficiency. J Inher Metab Dis 1997; 20:525–527.
286. Nishiyori A, Yoshino M, Kato H, Matsuura T, Hoshide R, Matsuda I, Kuno T, Miyazaki S, Hirose S-I, Kuromaru R-I, Mori M. The R40H mutation in a late onset type of human ornithine transcarbamylase deficiency in male patients. Hum Genet 1997; 99:171–176.
287. Ausems MGEM, Bakker E, Berger R, Duran M, van Diggelen OP, Keulemans JLM, de Valk HW, Kneppers ALJ, Dorland L, Eskes PF, Beemer FA, Poll-The BT, Smeitink JAM. Asymptomatic and late-onset ornithine transcarbamylase deficiency caused by a A208T mutation: clinical, biochemical and DNA analyses in a four-generation family. Amer J Med Genet 1997; 68:236–239.
288. Komaki S, Matsuura T, Oyanagi K, Hoshide R, Kiwaki K, Endo F, Shimadzu M, Matsuda I. Familial lethal inheritance of a mutated paternal gene in females causing X-linked ornithine transcarbamylase (OTC) deficiency. Amer J Med Genet 1997; 69:177–181.
289. Morizono H, Tuchman M, Rajagopal BS, McCann MT, Listrom CD, Yuan X, Venugopal D, Barany G, Allewell NM. Expression, purification and kinetic characterization of wild-type human ornithine transcarbamylase and a recurrent mutant that produces 'late onset' hyperammonaemia. Biochem J 1997; 322: 625–631.

290. Morizono H, Listrom CD, Rajagopal BS, Aoyagi M, McCann MT, Allewell NM, Tuchman M. 'Late onset' ornithine transcarbamylase deficiency: function of three purified recombinant mutant enzymes. Hum Molec Genet 1997; 6:963–968.
291. Oppliger Leibundgut E, Liechti-Gallati S, Colombo J-P, Wermuth B. Ornithine transcarbamylase deficiency: Ten new mutations and high proportion of de novo mutations in heterozygous females. Hum Mutation 1997; 9:409–411.
292. Matsuda I, Tanase S. The ornithine transcarbamylase (OTC) gene: Mutations in 50 Japanese families with OTC deficiency. Amer J Med Genet 1997; 71:378–383.
293. Keller C, Shapira SK, Clark GD. A urea cycle defect presenting as acute cerebellar ataxia in a 3-year-old girl. J Child Neurol 1998; 13:93–94.
294. Kogo T, Satoh Y, Kanazawa M, Yamamoto S, Takayanagi M, Ohtake A, Mori M, Niimi H. Expression analysis of two mutant human ornithine transcarbamylases in COS-7 cells. J Hum Genet 1998; 43:54–58.
295. Staudt M, Wermuth B, Freisinger P, Hassler A, Pontz BF. J Inher Metab Dis 1998; 21:71–72.
296. Yoo H-W, Kim G-H. Prenatal molecular evaluation of six fetuses in four unrelated Korean families with ornithine transcarbamylase deficiency. J Korean Med Sci 1998; 13:179–185.
297. Calvas P, Ségues B, Rozet J-M, Rabier D, Bonnefond J-P, Munnich A. Novel intragenic deletions and point mutations of the ornithine transcarbamylase gene in congenital hyperammonemia. Human Mutat 1998; Suppl 1: S81–S84.
298. Shimadzu M, Matsumoto H, Matsuura T, Kobayashi K, Komaki S, Kiwaki K, Hoshide R, Endo F, Saheki T, Matsuda I. Ten novel mutations of the ornithine transcarbamylase (OTC) gene in OTC deficiency. Human Mutat 1998; Suppl 1: S5–S7.
299. Popowska E, Ciara E, Rokicki D, Pronicka E. Three novel and one recurrent ornithine carbamoyltransferase gene mutations in Polish patients. J Inher Metab Dis 1999; 22:92–93.
300. Khoo ASB, Balraj P, Rachedi A, Chin CN, Volpi L. A novel complex mutation of the OTC (ornithine transcarbamylase)gene in a Malaysian pedigree. Hum Mutat, Mutat Polymorph 1999; Rep # 76 online.
301. Climent C, Garcia-Pérez MA, Sanjurjo P, Ruiz-Sanz J-I, Vilaseca MA, Pineda M, Campistol J, Rubio V. Identification of a cytogenetic deletion and of four novel mutations (Q69X, I172F, G188V, G197R) affecting the gene for ornithine transcarbamylase (OTC) in Spanish patients with OTC deficiency. Hum Mutat, Mutat in Brief 1999; #267 online.
302. McCullough BA, Yudkoff M, Batshaw ML, Wilson JM, Raper SE, Tuchman M. Genotype spectrum of ornithine transcarbamylase deficiency: correlation with the clinical and biochemical phenotype. Am J Med Genet 2000; 93:313–319.

303. Youssoufran H, Kazazian H, Phillips S, Aronis G, Tsiftisb, Brown V, Antonarakis S. Recurrent mutations in haemophilia A give evidence for CpG 12 mutation hotspots. Nature 1986; 324:380–382.
304. Plante RJ, Tuchman M. Polymorphisms in the human ornithine transcarbamylase gene useful for allele tracking. Hum Mutat, Mutat in Brief 1998; #193 online.
305. Cotton RGH, Rodrigues NR, Campbell RD. Reactivity of cytosine and thymine in single-base-pair mismatches with hydroxylamine and osmium tetroxide and its application to the study of mutations. Proc Natl Acad Sci USA 1988; 85:4397–4401.
306. Chou PY, Fasman GD. Empirical predictions of protein conformation. Ann Rev Biochem 1978; 47:251–276.
307. Kabsch W, Sander C. How good are predictions of protein secondary structure? FEBS Lett 1983; 155:179–182.
308. Presta LG, Rose GD. Helix signals in proteins. Science 1988; 240:1632–1641.
309. Richardson JS, Richardson DC. Amino acid preferences for specific locations at the ends of α helices. Science 1988; 240:1648–1652.
310. Branden C, Tooze J. Introduction to Protein Structure. 2nd Ed. New York, NY: Garland Publishing; 1999.
311. Legius E, Baten E, Stul M, Marynen P, Cassiman J-J. Sporadic late onset ornithine transcarbamylase deficiency in a boy with somatic mosaicism for an intragenic deletion. Clin Genet 1990; 38:155–159.
311a. Grompe M, Jones SN, Caskey CT. Molecular detection and correction of ornithine transcarbamylase deficiency. Trends in Genetics 1990; 6:335–339.

6. *Animal Models of OTC Deficiency and Their Gene Therapy*

SPARSE-FUR MUTANT MOUSE

312. Doolittle DP, Hulbert LL, Cordy C. A new allele of the sparse fur gene in the mouse. J Hered 1974; 65:194–195.
313. Briand P, Cathelineau L, Kamoun P, Gigot D, Penninckx M. Increase of ornithine transcarbamylase protein in sparse-fur mice with ornithine transcarbamylase deficiency. FEBS Lett 1981; 130:65–68.
314. Briand P, Cathelineau L. Sparse-fur mutation: a model for some human ornithine transcarbamylase deficiencies, Urea Cycle Diseases. Edited by A. Lowenthal, A. Mori, B. Marescau. New York, Plenum Press, 1982, pp. 185–194.
315. Qureshi IA, Letarte J, Ouellet R. Ornithine transcarbamylase deficiency in mutant mice. I. Studies on the characterization of enzyme defect and suitability as animal model of human disease. Pediat Res 1979; 13:807–811.
316. Briand P, Miura S, Mori M, Cathelineau L, Kamoun P, Tatibana M. Cell-

free synthesis and transport of precursors of mutant ornithine carbamoyltransferases into mitochondria. Biochim Biophys Acta 1983; 760:389–397.

317. Spector EB, Mazzocchi RA. The sparse fur mouse: an animal model for a human inborn error of metabolism of the urea cycle. Prog Clin Biol Res 1983; 127: 85–96.
318. Ohtake A, Takayanagi M, Yamamoto S, Kakinuma H, Nakajima H, Tatibana M, Mori M. Molecular basis of ornithine transcarbamylase deficiency in spf and spf-ash mutant mice. J Inher Metab Dis 1986; 9:289–291.
319. Qureshi IA, Letarte J, Ouellet R. Spontaneous animal models of ornithine transcarbamylase deficiency: studies on serum and urinary nitrogenous metabolites, Urea Cycle Diseases. Edited by A. Lowenthal, A. Mori, B. Marescau. New York, Plenum Press, 1982, pp. 173–183.
320. Qureshi IA, Letart J, Quellet R. Expression of ornithine transcarbamylase deficiency in the small intestine and colon of sparse-fur mutant mice. J Pediat Gastroent Nutr 1985; 4:118–124.
321. Dubois N, Cavard C, Chasse J-F, Kamoun P, Briand P. Compared expression levels of ornithine transcarbamylase and carbamylphosphate synthetase in liver and small intestine of normal and mutant mice. Biochim Biophys Acta 1988; 950: 321–328.
322. Qureshi IA, Rama Rao KV. Sparse-fur (spf) mouse as a model of hyperammonemia: alterations in the neurotransmitter systems, in Advances in Cirrhosis, Hyperammonemia, and Hepatic Encephalopathy. Edited by V Felipo, S Grisolia. Adv Exper Med Biol 1997; 420: 143–158.
323. Robinson MB, Hopkins K, Batshaw ML, McLaughlin BA, Heyes MP, Oster-Granite ML. Evidence of excitotoxicity in the brain of the ornithine carbamoyltransferase deficient *sparse fur* mouse. Develop Brain Res 1995; 90: 35–44.
324. Qureshi K, Rama Rao KV, Qureshi IA. Differential inhibition by hyperammonemia of the electron transport chain enzymes in synaptosomes and non-synaptic mitochondria in ornithine transcarbamylase-deficient spf-mice: restoration by acetyl-L-carnitine. Neurochem Res 1998; 23: 855–861.
325. Hopkins KJ, Oster-Granite ML. Characterization of N-methyl-D-aspartate receptors in the hyperammonemic *Sparse fur* mouse. Brain Res 1998; 797: 209–217.
326. Hopkins KJ, McKean J, Mervis RF, Oster-Granite ML. Dendritic alterations in cortical pyramidal cells in the *sparse fur* mouse. Brain Res 1998; 797: 167–172.
327. Windmueller HG, Spaeth AE. Source and fate of circulating citrulline. Am J Physiol 1981; 241: E473–E480.
328. Ratnakumari L, Qureshi IA, Butterworth RF. Effect of L-carnitine on cere-

bral and hepatic energy metabolites in congenitally hyperammonemic sparse-fur mice and its role during benzoate therapy. Metabolism 1993; 42: 1039–1046.

329. Rao KVR, Qureshi IA. Reduction in the MK-801 binding sites of the NMDA sub-type of glutamate receptor in a mouse model of congenital hyperammonemia: prevention by acetyl-L-carnitine. Neuropharmacol 1999; 38: 383–394.
330. Butterworth RF. Evidence for a central cholinergic deficit in congenital ornithine transcarbamylase deficiency. Dev Neurosci 1998: 20: 478–484.
331. Butterworth RF. Effects of hyperammonaemia on brain function. J Inher Metab Dis 1998; 21 (Suppl 1): 6–20.
332. Albrecht J, Jones EA. Hepatic encephalopathy: molecular mechanisms underlying the clinical syndrome. J Neurol Sci 1999; 170: 138–146.
333. Nelson J, Qureshi IA, Vasudevan S, Sarma DSR. The effects of various inhibitors on the regulation of orotic acid excretion in sparse-fur mutant mice (spf/Y) deficient in ornithine transcarbamylase. Chem Biological Interactions 1993; 89: 35–47.
334. Seiler N, Grauffel C, Daune-Anglard G, Sarhan S, Knödgen B. Decreased hyperammonaemia and orotic aciduria due to inactivation of ornithine aminotransferase in mice with a hereditary abnormal ornithine carbamoyltransferase. J Inher Metab Dis 1994: 17: 691–703.
335. Batshaw ML, Hyman SL, Coyle JT, Robinson MB, Qureshi IA, Mellits ED, Quaskey S. Effect of sodium benzoate and sodium phenylacetate on brain serotonin turnover in the ornithine transcarbamylase-deficient sparse-fur mouse. Pediatr Res 1988; 23: 368–374.

SPARSE FUR^{ASH} MUTANT MOUSE

336. Rosenberg LE, Kalousek F, Orsulak MD. Biogenesis of ornithine transcarbamylase in spf^{ash} mutant mice: two cytoplasmic precursors, one mitochondrial enzyme. Science 1983; 222:426–428.
337. Hodges PE, Rosenberg LE. The spf^{ash} mouse: a missense mutation in the ornithine transcarbamylase gene also causes aberrant mRNA splicing. Proc Natl Acad Sci USA 1989; 86:4142–4146.
338. Briand P, Baudouin F, Rabier D, Cathelineau L. Ornithine transcarbamylase deficiencies in human males. Kinetic and immunochemical classification. Biochim Biophys Acta 1982; 704:100–106.
339. Cohen NS, Cheung C-W, Raijman L. Altered enzyme activities and citrulline synthesis in liver mitochondria from ornithine carbamoyltransferase-deficient *sparse-fur*^{ash} mice. Biochem J 1989; 257:251–257.
340. Snodgrass PJ. Unpublished observations, 1995.
341. Schimke RT. Adaptive characteristics of urea cycle enzymes in the rat. J Biol Chem 1962; 237: 459–468.
342. Snodgrass PJ, Lin RC. Induction of urea cycle enzymes of rat liver by amino acids. J Nutr 1981; 111: 586–601.

343. Snodgrass PJ. Dexamethasone and glucagon cause synergistic increases of urea cycle enzyme activities in livers of normal but not adrenalectomized rats. Enzyme 1991; 45: 30–38.
344. Mori M, Miura S. Tatibana M, Cohen PP. Cell-free translation of carbamyl phosphate synthetase 1 and ornithine transcarbamylase messenger RNAs of rat liver. J Biol Chem 1981; 256: 4127–4132.
345. Cavard C, Grimber G, Dubois N, Chasse J-F, Bennoun M, Minet-Thuriaux M, Kamoun P, Briand P. Correction of mouse ornithine transcarbamylase deficiency by gene transfer into the germ line. Nucleic Acids Res 1998; 16: 2099–2110.
346. Murakami T, Takiguchi M, Inomoto T, Yamamura K-I, Mori M. Tissue- and developmental stage-specific expression of the rat ornithine carbamoyltransferase gene in transgenic mice. Developmental Genetics 1989; 10: 393–401.
347. Jones SN, Grompe M, Idrees Munir M, Veres G, Craigen WJ, Caskey CT. Ectopic correction of ornithine transcarbamylase deficiency in sparse fur mice. J Biol Chem 1990; 265: 14684–14690.
348. Shimada T, Noda T, Tashiro M, Murakami T, Takiguchi M, Mori M, Yamamura K-I, Saheki T. Correction of ornithine transcarbamylase (OTC) deficiency in spf-ash mice by introduction of rat OTC gene. FEBS Lett 1991; 279: 198–200.
349. Saheki T, Mori K, Kobayashi K, Horiuchi M, Shige T, Obara T, Suzuki S, Mori M, Yamamura K-I. Importance of ornithine transcarbamylase (OTC) deficiency in small intestine for urinary orotic acid excretion: analysis of OTC-deficient spf-ash mice with OTC transgene. Biochim Biophys Acta 1995; 1270: 87–93.
350. Ledley FD. Hepatic gene therapy: present and future. Hepatol 1993; 18: 1263–1273.
351. Chang AGY, Wu GY. Gene therapy: applications to the treatment of gastrointestinal and liver diseases. Gastroent 1994; 106: 1076–1084.
352. Stratford-Perricaudet LD, Levrero M, Chasse J-F, Perricaudet M, Briand P. Evaluation of the transfer and expression in mice of an enzyme-encoding gene using a human adenovirus vector. Human Gene Therapy 1990; 1: 241–256.
353. Morsy MA, Alford EL, Bett A, Graham FL, Caskey CT. Efficient adenoviral-mediated ornithine transcarbamylase expression in deficient mouse and human hepatocytes. J Clin Invest 1993; 92: 1580–1586.
354. Cristiano RJ, Smith LC, Woo SLC. Hepatic gene therapy: adenovirus enhancement of receptor-mediated gene delivery and expression in primary hepatocytes. Proc Natl Acad Sci USA 1993; 90: 2122–2126.
355. Morsy MA, Caskey CT. Ornithine transcarbamylase deficiency: a model for gene therapy, in Advances in Hepatic Encephalopathy, Hyperammonemia and Ammonia Toxicity. Ed. by V Felipo, S Grisolia. Adv Exper Med Biol 1994; 368:145–154.

356. Yang Y, Nunes FA., Berencsi K, Furth EE, Gönczöl E, Wilson JM. Cellular immunity to viral antigens limits E1-deleted adenoviruses for gene therapy. Proc Natl Acad Sci USA 1994; 91: 4407–4411.
357. Ye X, Robinson MB, Batshaw ML, Furth EE, Smith I, Wilson JM. Prolonged metabolic correction in adult ornithine transcarbamylase-deficient mice with adenoviral vectors. J Biol Chem 1996; 271: 3639–3646.
358. Morsy MA, Zhao JZ, Ngo TT, Warman AW, O'Brien WE, Graham FL, Caskey CT. Patient selection may affect gene therapy success. Dominant negative effects observed for ornithine transcarbamylase in mouse and human hepatocytes. J Clin Invest 1996; 97: 826–832.
359. Wente SR, Schachman HK. Shared active sites in oligomeric enzymes: model studies with defective mutants of aspartate transcarbamoylase produced by site directed mutagenesis. Proc Natl Acad Sci USA 1987; 84: 31–35.
360. Kiwaki K, Kanegae Y, Saito I, Komaki S, Nakamura K, Miyazaki J-I, Endo F, Matsuda I. Correction of ornithine transcarbamylase deficiency in adult *spf*^{ash} mice and in OTC-deficient human hepatocytes with recombinant adenoviruses bearing the CAG promoter. Human Gene Therapy 1996; 7: 821–830.
361. Ye X, Robinson MB, Pabin C, Quinn T, Jawad A, Wilson JM, Batshaw ML. Adenovirus-mediated *in vivo* gene transfer rapidly protects ornithine transcarbamylase-deficient mice from an ammonium challenge. Pediat Res 1997; 41: 527–534.
362. Batshaw ML, Robinson MB, Ye X, Pabin C, Daikhin Y, Burton BK, Wilson JM, Yudkoff M. Correction of ureagenesis after gene transfer in an animal model and after liver transplantation in humans with ornithine transcarbamylase deficiency. Pediat Res 1999; 46: 588–593.
363. Ye X, Robinson MB, Pabin C, Batshaw ML, Wilson JM. Transient depletion of CD4 lymphocyte improves efficacy of repeated administration of recombinant adenovirus in the ornithine transcarbamylase deficient sparse fur mouse. Gene Therapy 2000; 7:1761–1767.
364. Raper SE, Wilson JM, Yudkoff M, Robinson MB, Ye X, Batshaw ML. Developing adenoviral-mediated *in vivo* gene therapy for ornithine transcarbamylase deficiency. J Inher Metab Dis 1998; 21(Suppl 1): 119–137.
365. Grompe M, Jones SN, Loulseged H, Caskey CT. Retroviral-mediated gene transfer of human ornithine transcarbamylase into primary hepatocytes of *spf* and *spf-ash* mice. Human Gene Therapy 1992; 3: 35–44.
366. Podevin G, Ferry M, Calise D, Revillon Y. In vivo retroviral-mediated transfer of a marker-gene in ornithine transcarbamylase-deficient spf^{ash} mice. J Pediat Surg 1996; 31: 1516–1519.
367. Saheki T, Tomomura M, Horiuchi M, Imamura Y, Tomomura A, Abu Musa DMA, Kobayashi K. Abnormal gene expression causing hyperammonemia in carnitine-deficient juvenile visceral steatosis (JVS) mice , in

Advances in Cirrhosis, Hyperammonemia, and Hepatic Encephalopathy, ed. by V. Felipo, S. Grisola. Adv Exper Med Biol 1997; 420: 159–170.

368. Yokogawa K, Yonekawa M, Tamai I, Ohashi R, Tatsumi Y, HigashiY, Nomura M, Hashimoto N, Nikaido H, Hayakawa J-I, Nezu J-I, Oku A, Shimane M, Miyamoto K-I, Tsuji A. Loss of wild-type carrier-mediated L-carnitine transport activity in hepatocytes of Juvenile Visceral Steatosis mice. Hepatol 1999; 30:997–1001.

7. *Clinical and Laboratory Findings in OTC Deficiency*

369. Russell A, Levin B, Oberholzer VG, Sinclair L. Hyperammonemia: a new instance of an inborn enzymatic defect of the biosynthesis of urea. Lancet 1962; 2:699–700.

370. Levin B, Russell A. Treatment of Hyperammonemia. Amer J Dis Child 1967; 113:142–145.

371. Levin B. Hyperammonemia. An inherited disorder of urea biosynthesis due to liver ornithine transcarbamylase deficiency. Sixth Annual Symposium of the Society for Study of Inborn Errors of Metabolism, Zurich 1968b; pp. 123–130.

372. Levin B, Oberholzer VG, Sinclair L. Biochemical investigations of hyperammonaemia. Lancet 1969; 2:170–174.

373. Levin B, Abraham JM, Oberholzer VG, Burgess EA. Hyperammonaemia: a deficiency of liver ornithine transcarbamylase. Occurrence in mother and child. Arch Dis Childh 1969; 44:152–161.

374. Abraham JM, Levin B. Hyperammonaemia in mother and child. Sixth Annual Symposium of the Society for Study of Inborn Errors of Metabolism, Zurich 1968; pp. 131–133.

375. Hopkins IJ, Connelly JF, Dawson AG, Hird FJR, Maddison TG. Hyperammonemia due to ornithine transcarbamylase deficiency. Arch Dis Childh 1969; 44: 143–148.

376. Herrin JT, McCredie DA. Peritoneal dialysis in the reduction of blood ammonia levels in a case of hyperammonaemia. Arch Dis Childh 1969; 44:149–151.

377. Corbeel L, Colombo JP, van Sande M, Weber A. Periodic attacks of lethargy in a baby with ammonia intoxication and defect in ureogenesis. Sixth Annual Symposium of the Society for Study of Inborn Errors of Metabolism, Zurich 1968; pp. 115–118.

378. Corbeel LM, Colombo JP, van Sande M, Weber A. Periodic attacks of lethargy in a baby with ammonia intoxication due to a congenital defect in ureogenesis. Arch Dis Childh 1969; 44:681–687.

379. Nagayama E, Kitayama T, Oguchi H, Ogata K, Tamura E, Onisawa J. Hyperammonemia: a deficiency of liver ornithine transcarbamylase. Paediatrics Univ Tokyo 1970; 18:167–173.

380. Matsuda I, Arashima S, Nambu H, Takekoshi Y, Anakura M. Hyperammonemia due to a mutant enzyme of ornithine transcarbamylase. Pediatrics 1971; 48: 595–600.
381. Sunshine P, Lindenbaum JE, Levy HL, Freeman JM. Hyperammonemia due to a defect in hepatic ornithine transcarbamylase. Pediatrics 1972; 50:100–111.
382. Cathelineau L, Navarro J, Aymard P, Baudon J-J, Mondet Y, Polonovski C, Laplane R. Hyperammoniemie hereditaire par anomalie qualitative de l'ornithine-carbamyl-transferase hepatique et intestinale. Arch Franc Ped 1972; 29:713–736.
383. Salle B, Levin B, Longin B, Richard P, Andre M, Gauthier J. Hyperammoniemie congenitale par deficit en ornithine carbamyl transferase et carbamyl phosphate synthetase. Arch Franc Ped 1972; 29:493–504.
384. Short EM, Conn HO, Snodgrass PJ, Campbell AGM, Rosenberg LE. Evidence for X-linked dominant inheritance of ornithine transcarbamylase deficiency. N Engl J Med 1973; 288:7–12.
385. Krebs HA, Hems R, Lund P. Regulatory mechanisms in the synthesis of urea, Inborn Errors of Metabolism. Edited by FH Hommes, van den Berg CJ, London, Academic Press, 1973, pp. 201–219.
386. Morley C, Sardharwalla IB. Case of hyperammonaemia due to ornithine transcarbamylase deficiency. Arch Dis Childh 1974; 49:747.
387. Palmer T, Oberholzer VG, Burgess EA, Butler LJ, Levin B. Hyperammonaemia in 20 families. Biochemical and genetical survey, including investigations in 3 new families. Arch Dis Childh 1974; 49:443–449.
388. Beaudry MA, Letarte J, Collu R, Leboeuf G,, Ducharme JR, Melançon SB, Dallaire L. Hyperammoniemie chronique avec acidurie orotique: stimulation de la voie des pyrimidines. Diabete & Metabolisme (Paris) 1975; 1:29–37.
389. Qureshi IA, Letarte J, Ouellet R. Study of enzyme defect in a case of ornithine transcarbamylase deficiency. Diabete & Metabolisme (Paris) 1978; 4:239–241.
390. Gray RGF, Black JA, Lyons VH, Pollitt RJ. Ornithine transcarbamylase deficiency: enzyme studies on a further case and a method of diagnosis using plasma enzyme ratios. Pediatr Res 1976; 10:918–923.
391. Krieger I, Bachmann C, Gronemeyer WH, Cejka J. Propionic acidemia and hyperlysinemia in a case with ornithine transcarbamylase (OTC) deficiency. J Clin Endocrinol Metab 1976; 43:796–802.
392. Glasgow AM, Kraegel JH, Schulman JD. Studies of the cause and treatment of hyperammonemia in females with ornithine transcarbamylase deficiency. Pediatrics 1978; 62:30–37.
393. Hokanson JT, O'Brien WE, Idemoto J, Schafer IA. Carrier detection in ornithine transcarbamylase deficiency. J Pediatr 1978; 93:75–78.
394. McReynolds JW, Mantagos S, Brusilow S, Rosenberg LE. Treatment of

complete ornithine transcarbamylase deficiency with nitrogen-free analogues of essential amino acids. J Pediatr 1978; 93:421–427.

395. LaBrecque DR, Latham PS, Riely CA, Hsia YE,, Klatskin G. Heritable urea cycle enzyme deficiency-liver disease in 16 patients. J Pediatr 1979; 94:580–587.
396. Heick HMC, Shipman RT, Norman MG, James W. Reye-like syndrome associated with use of insect repellent in a presumed heterozygote for ornithine carbamoyl transferase deficiency. J Pediatr 1980; 97:471–473.
397. Shapiro JM, Schaffner F, Tallan HH, Gaull GE. Mitochondrial abnormalities of liver in primary ornithine transcarbamylase deficiency. Pediatr Res 1980; 14:735–739.
398. Kodama H, Nose O, Okada S, Yabuuchi H. The study of organic acids metabolism in a patient with ornithine transcarbamylase (OTC) deficiency. Adv Exper Med Biol 1982; 153:341–350.
399. Kodama H, Samukawa K, Okada S, Nose O, Maki I, Yamaguchi M, Yabuuchi H. Study of ammonia metabolism in a patient with ornithine transcarbamylase deficiency using an ^{15}N tracer. Clin Chim Acta 1983; 132:267–275.
400. Kodama H, Nagayama H, Shimoizumi H, Okabe I, Kamoshita S. Immunochemical study of ornithine transcarbamylase deficiency. J Inher Metab Dis 1984; 7:131–132.
401. Kodama H, Ohtake A, Mori M, Okabe I, Tatibana M, Kamoshita S. Ornithine transcarbamylase deficiency:a case with a truncated enzyme precursor and a case with undetectable mRNA activity. J Inher Metab Dis 1986; 9:175–185.
402. Kendall BE, Kingsley DPE, Leonard JV, Lingam S, Oberholzer VG. Neurological features and computed tomography of the brain in children with ornithine carbamoyl transferase deficiency. J Neurol Neurosurg Psych 1983; 46:28–34.
403. Takayanagi M, Ohtake A, Ogura N, Nakajima H, Hoshino M. A female case of ornithine transcarbamylase deficiency with marked computed tomographic abnormalities of the brain. Brain Dev 1984; 6:58–60.
404. Matsuda I, Nagata N, Ohyanagi K, Tsuchiyama A, Yamamoto H, Hase Y, Kodama H, Kai Y. Biochemical heterogeneity of ornithine carbamoyl transferase(OCT) in patients with OCT deficiency. Jpn J Human Genet 1984; 29:327–333.
405. Woodfin BM, Davis LE, Bernard LR, Kornfeld M. A fatal variant of human ornithine carbamoyl transferase is stimulated by $Mg2+$. Biochem Med Metab Biol 1986; 36:300–305.
406. Rowe PC, Newman SL, Brusilow SW. Natural history of symptomatic partial ornithine transcarbamylase deficiency. N Engl J Med 1986; 314:541–547.
407. Batshaw ML, Msall M, Beaudet AL, Trojak J. Risk of serious illness in het-

erozygotes for ornithine transcarbamylase deficiency. J Pediatr 1986; 108:236–241.

408. Lacey DJ, Duffner PK, Cohen ME, Mosovich L. Unusual biochemical and clinical features in a girl with ornithine transcarbamylase deficiency. Pediatr Neurol 1986; 2:51–53.
409. Gilchrist JM, Coleman RA. Ornithine transcarbamylase deficiency: adult onset of severe symptoms. Ann Int Med 1987; 106:556–558.
410. Hayasaka K, Metoki K, Ishiguro S, Kato S, Chiba T, Hirooka M, Kikuchi M, Kurobane I, Narisawa K, Tada K. Partial ornithine transcarbamylase deficiency in females: diagnosis by an immunohistochemical method. Eur J Pediatr 1987; 146:370–372.
411. Girgis N, McGravey V, Shah BL, Herrin J, Shih VE. Lethal ornithine transcarbamylase deficiency in a female neonate. J Inher Metab Dis 1987; 10:274–275.
412. Rowe PC, Valle D, Brusilow SW. Inborn errors of metabolism in children referred with Reye's syndrome.? A changing pattern. JAMA 1988; 260:3167–3170.
413. Wendel U, Wilichowski E, Schmidtke J, Bachmann C. DNA analysis of ornithine transcarbamylase deficiency. Eur J Pediatr 1988; 147:368–371.
414. Largillière C, Houssin D, Gottrand F, Mathey C, Checoury A, Alagille D, Farriaux J-P. Liver transplantation for ornithine transcarbamylase deficiency in a girl. J Pediatr 1989; 115:415–417.
415. Aida S, Ogata T, Kamota T, Nakamura N. Primary ornithine transcarbamylase deficiency. A case report and electron microscopic study. Acta Pathol Jpn 1989; 39:451–456.
416. de Grauw TJ, Smit LME,, Brockstedt M, Meijer Y, van der Klei-van Moorsel J, Jakobs C. Acute hemiparesis as the presenting sign in a heterozygote for ornithine transcarbamylase deficiency. Neuropediatr 1990; 21:133–135.
417. Arn PH, Hauser ER, Thomas GH, Herman G, Hess D, Brusilow SW. Hyperammonemia in women with a mutation at the ornithine carbamoyltransferase locus. A cause of postpartum coma. N Engl J Med 1990; 322:1652–1655.
418. Soulpis C, Markosoglou D, Papadelis F, Caraboula H, Giouroukos S, Skarpalezou A, Missiou-Tsagarakis S, Michelakakis H. Ornithine transcarbamylase deficiency: findings and treatment in a symptomatic female heterozygote. J Inher Metab Dis 1991; 14:107–108.
419. Rabier D, Narcy C, Bardet J, Parvy P, Saudubray JM, Kamoun P. Arginine remains an essential amino acid after liver transplantation in urea cycle enzyme deficiencies. J Inher Metab Dis 1991; 14:277–280.
420. Tokatli A, Coşkun T, Çataltepe Ş, Özalp I. Valproate-induced lethal hyperammonaemic coma in a carrier of ornithine carbamoyltransferase deficiency. J Inher Metab Dis 1991; 14:836–837.
421. Berrez J-M, Bardot O, Thiard M-C, Alvarez F, Latruffe N. Molecular anal-

ysis of a human liver mitochondrial ornithine transcarbamylase deficiency. J Inher Metab Dis 1991; 14:29–36.
422. Mamourian AC, du Plessis A. Urea cycle defect: a case with MR and CT findings resembling infarct. Pediatr Radiol 1991; 21:594–595.
423. László A, Karsai T, Várkonyi Á. Congenital hyperammonemia: symptomatic carrier girl patient and her asymptomatic heterozygous mother for ornithine transcarbamylase (OTC) deficiency: specific enzyme diagnostic and kinetic investigations for the detection of heterozygous genostatus. Acta Paediatr Hungarica 1991; 31:291–297.
424. Christodoulou J, Qureshi IA, McInnes RR, Clarke JTR. Ornithine transcarbamylase deficiency presenting with strokelike episodes. J Pediatr 1993; 122:423–425.
425. Connelly A, Cross JH, Gadian DG, Hunter JV, Kirkham FJ, Leonard JV. Magnetic resonance spectroscopy shows increased brain glutamine in ornithine carbamoyl transferase deficiency. Pediatr Res 1993; 33:77–81.
426. Perini M, Zarcone D, Corbetta C. Hyperammoniemic coma in an adolescent girl: an unusual case of ornithine transcarbamylase deficiency. Ital J Neurol Sci 1993; 14:461–464.
427. Fries MH, Kuller JA, Jurecki E, Packman S. Prenatal counseling in heterozygotes for ornithine transcarbamylase deficiency. Clin Pediatr 1994; 33:525–529.
428. Lerman-Sagie T, Mimouni M. Reversal of anorexia in a child with partial ornithine transcarbamylase deficiency by cyproheptadine therapy. Clin Pediatr 1995; 34:163–165.
429. Largillière C. Psychiatric manifestations in girl with ornithine transcarbamylase deficiency. Lancet 1995; 345:1113.
430. Pridmore CL, Clarke JTR, Blaser S. Ornithine transcarbamylase deficiency in females: an often overlooked cause of treatable encephalopathy. J Child Neurol 1995; 10:369–374.
431. Felig DM, Brusilow SW, Boyer JL. Hyperammonemic coma due to parenteral nutrition in a woman with heterozygous ornithine transcarbamylase deficiency. Gastroenterology 1995; 109:282–284.
432. Mattson LR, Lindor NM, Goldman DH, Goodwin JT, Groover RV, Vockley J. Central pontine myelinolysis as a complication of partial ornithine carbamoyl transferase deficiency. Am J Med Genet (Neuropsych Genet) 1995; 60: 210–213.
433. Hasegawa T, Tzakis AG, Todo S, Reyes J, Nour B, Finegold DN, Starzl TE. Orthotopic liver transplantation for ornithine transcarbamylase deficiency with hyperammonemic encephalopathy. J Pediatr Surg 1995; 30:863–865.
434. Leão M. Valproate as a cause of hyperammonemia in heterozygotes with ornithine-transcarbamylase deficiency. Neurol 1995; 45:593–594.
435. Bajaj SK, Kurlemann G, Schuierer G, Peters PE. CT and MRI in a girl with late-onset ornithine transcarbamylase deficiency: case report. Neuroradiol 1998; 38:796–799.

436. Inui A, Fujisawa T, Komatsu H, Tanaka K, Inui M. Histological improvement in native liver after auxiliary partial liver transplantation for ornithine transcarbamylase deficiency. Lancet 1996; 348:751–752.
437. Egawa H, Tanaka K, Inomata Y, Uemoto S, Okajima H, Satomura K, Kiuchi T, Yabe S, Nishizawa H, Yamaoka Y. Auxiliary partial orthotopic liver transplantation from a living related donor: a report of two cases. Transplant Proc 1996; 28:1071–1072.
438. Kasahara M, Kiuchi T, Uryuhara K, Ogura Y, Takakura K, Egawa H, Asonuma K, Uemoto S, Inomata Y, Tanake K. Treatment of ornithine transcarbamylase deficiency in girls by auxiliary liver transplantation: conceptual changes in a living-donor program. J Pediatr Surg 1998; 33:1753–1756.
439. Demmer LA, Kim JM, de Martinville B, Dowton SB. A novel missense mutation in the exon containing the putative ornithine-binding domain of the OTC enzyme in a female. Hum Mutation 1996; 7:279.
440. Zammarchi E, Donati MA, Filippi L, Resti M. Cryptogenic hepatitis masking the diagnosis of ornithine transcarbamylase deficiency. J Pediatr Gastroenterol Nutrition 1996; 22:380–383.
441. Yeh S-K, Hwu W-L, Tsai W-S Wu T-J, Tuchman M, Wang T-R. Ornithine transcarbamylase deficiency. J Formos Med Assoc 1997; 96:43–45.
442. Komaki S, Matsuura T, Oyanagi K, Hoshide R, Kiwaki K, Endo F, Shimadzu M, Matsuda I. Familial lethal inheritance of a mutated paternal gene in females causing X-linked ornithine transcarbamylase (OTC) deficiency. Am J Med Genet 1997; 69:177–181.
443. Uemoto S, Yabe S, Inomata Y, Nishizawa H, Asonuma K, Egawa H, Kiuchi T, Okajima H, Yamaoka Y, Yamabe H, Inui A, Fujisawa T, Tanaka K. Coexistence of a graft with the preserved native liver in auxiliary partial orthotopic liver transplantation from a living donor for ornithine transcarbamylase deficiency. Transplantation 1997; 63:1026–1028.
444. Heringlake S, Böker K, Manns M. Fatal clinical course of ornithine transcarbamylase deficiency in an adult heterozygous female patient. Digestion 1997; 58:83–86.
445. Klosowski S, Largillière C, Storme L, Rakza T, Rabier D, Lequien P. Lethal ornithine transcarbamylase deficiency in a female neonate: a new case. Acta Paediatr 1998; 87:227–230.
446. Oechsner M, Steen C, Stürenburg HJ, Kohlschütter A. Hyperammonaemic encephalopathy after initiation of valproate therapy in unrecognized ornithine transcarbamylase deficiency. J Neurol Neurosurg Psychiatry 1998; 64:680–682.
447. Martland T, Mbamali AC, Rittey C, Tanner S, Bonham JR, Griffiths PD. Ornithine transcarbamylase deficiency: a case report. Neuropediatr 1998; 29:331–332.
448. Schwab S, Schwarz S, Mayatepek E, Hoffmann GF. Recurrent brain edema in ornithine-transcarbamylase deficiency. J Neurol 1999; 246:609–611.
449. Cathelineau L, Saudubray JM, Navarro J, Polonovski C. Transmission par

le chromosome X du gène de structure de l'ornithine-carbamyl-transférase. Étude de trois familles. Ann Génét 1973; 16:173–182.
450. Goldstein AS, Hoogenraad NJ, Johnson JD, Fukanaga K, Swierczewski E, Cann HM, Sunshine P. Metabolic and genetic studies of a family with ornithine transcarbamylase deficiency. Pediat Res 1974; 8:5–12.
451. Farriaux JP, Dhondt JL, Cathelineau L, Ratel J, Fontaine G. Hyperammonemia through deficiency of ornithine carbamyl transferase. Z Kinderheilk 1974; 118:231–247.
452. Saudubray JM, Cathelineau L, Laugier JM, Charpentier C, Lejeune JA, Mozziconacci P. Hereditary ornithine transcarbamylase deficiency. Report of two male cases with residual enzymatic activity. Acta Paediatr Scand 1975; 64:464–472.
453. Amir J, Alpert G, Statter M, Gutman A, Reisner SH. Intracranial haemorrhage in siblings and ornithine transcarbamylase deficiency. Acta Paediatr Scand 1982; 71:671–673.
454. Augustin L, Mavinakere M, Morizono H, Tuchman M. Expression of wild-type and mutant human ornithine transcarbamylase genes in Chinese hamster ovary cells and lack of dominant negative effect of R141Q and R40H mutants. Pediatr Res 2000; 48:842–846.
455. Rottem M, Statter M, Amit R, Brand N, Bujanover Y, Yatziv S. Clinical and laboratory study in 22 patients with inherited hyperammonemic syndromes. Isr J Med Sci 1986; 22:833–836.
456. Liechti S, Vici CD, Bachmann C, Mazziotta MRM, Bartuli A, Sabetta G. Prenatal exclusion of ornithine transcarbamylase (OTC) by using RFLP analysis. J Inher Metab Dis 1990; 13:888–890.
457. Pelet A, Rotig A, Bonaiti-Pellié C, Rabier D, Cormier V, Toumas E, Hentzen D, Saudubray J-M, Munnich A. Carrier detection in a partially dominant X-linked disease: ornithine transcarbamylase deficiency. Hum Genet 1990; 84:167–171.
458. Yoshino M, Nishiyori J, Yamashita F, Kumashiro R, Abe H, Tanikawa K, Ohno T, Nakao K, Kaku N, Fukushima H, Kubota K. Ornithine transcarbamylase deficiency in male adolescence and adulthood. Enzyme 1990; 43:160–168.
459. Ahrens MJ, Berry SA, Whitley CB, Markowitz DJ, Plante RJ, Tuchman M. Clinical and biochemical heterogeneity in females of a large pedigree with ornithine transcarbamylase deficiency due to the R141Q mutation. Am J Med Genet 1996; 66:311–315.
460. Segues B, Rozet J-M, Gilbert B, Saugier-Veber P, Rabier D, Saudubray J-M, Carré M, Rouleau FP, Menget A, Bonardi J-M, Lyonnet S, Bonnefont J-P, Munnich A. Apparent segregation of null alleles ascribed to deletions of the ornithine transcarbamylase gene in congenital hyperammonaemia. Prenatal Diagn 1995; 15:757–761.
461. Hou J-W, Wang T-R. Amino acid and DNA analyses in a family with ornithine transcarbamylase deficiency. J Formos Med Assoc 1996; 95:144–147.

462. Topaloglu AK, Sansaricq C, Fox JE, Bale AE, Tuchman M, Desnick RJ. Prenatal molecular diagnosis of severe ornithine carbamoyltransferase deficiency due to a novel mutation, E181G. J Inher Metab Dis 1999; 22:82–83.
463. Shih VE,, Safran AP, Ropper AH, Tuchman M. Ornithine carbamoyltransferasse deficiency: unusual clinical findings and novel mutation. J Inher Metab Dis 1999; 22:672–673.
464. Bowling F, McGown I, McGill J, Cowley D, Tuchman M. Maternal gonadal mosaicism causing ornithine transcarbamylase deficiency. Am J Med Genet 1999; 85:452–454.
465. Maestri NE, Brusilow SW, Clissold DB, Bassett SS. Long-term treatment of girls with ornithine transcarbamylase deficiency. N Engl J Med 1996; 335:855–859.
466. Batshaw ML, Roan Y, Jung AL, Rosenberg LA, Brusilow SW. Cerebral dysfunction in asymptomatic carriers of ornithine transcarbamylase deficiency. N Engl J Med 1980; 302:482–485.
467. Campbell AGM, Rosenberg LE, Snodgrass PJ, Nuzum CT. Lethal neonatal hyperammonemia due to complete ornithine-transcarbamylase deficiency. Lancet 1971; 2:217–218.
468. Campbell AGM, Rosenberg LE, Snodgrass PJ, Nuzum CT. Ornithine transcarbamylase deficiency. A cause of lethal neonatal hyperammonemia in males. N Engl J Med 1973; 288:1–6.
469. Scott CR, Teng CC, Goodman SI, Greensher A, Mace JW. X-linked transmission of ornithine-transcarbamylase deficiency. Lancet 1972; 2:1148.
470. Kang ES, Snodgrass PJ, Gerald PS. Ornithine transcarbamylase deficiency in the newborn infant. J Pediatr 1973; 82:642–649.
471. Saudubray J-M, Cathelineau L, Charpentier C, Boisse J, Allaneau C, LeBont H, Lesage B. Deficit hereditaire en ornithine-carbamyl-transferase avec anomalie enzymatique qualitative. Arch Franç Péd 1973; 30:15–27.
472. Gelehrter TD, Rosenberg LE. Ornithine transcarbamylase deficiency. Unsuccessful therapy of neonatal hyperammonemia with N-carbamyl-L-glutamate and L-arginine. N Engl J Med 1975; 292:351–352.
473. Snyderman SE, Sansaricq C, Phansalkar SV, Schacht RG, Norton PM. The therapy of hyperammonemia due to ornithine transcarbamylase deficiency in a male neonate. Pediatr 1975; 56:65–73.
474. Stoll C, Bieth R, Dreyfus J, Flori E, Lutz P, Levy J-M. Une nouvelle famillle avec mutation-du gene de structure de l'ornithine carbamyltransferase humaine. Arch Franç Péd 1978; 35:512–518.
475. Batshaw ML, Painter MJ, Sproul GT, Schafer IA, Thomas GH, Brusilow S. Therapy of urea cycle enzymopathies: three case studies. The J Hopkins Med J 1981; 148:34–40.
476. Michels VV, Potts E, Walser M, Beaudet AL. Ornithine transcarbamylase deficiency: long-term survival. Clin Genet 1982; 22:211–214.
477. Brubakk A-M, Teijema LL, Blom W, Berger R. Successful treatment of severe OTC deficiency. J Pediatr 1982; 100:929–931.

478. Guibaud P, Baxter P, Bourgeois J, Louis JJ, Bureau J. Severe ornithine transcarbamylase deficiency. Two and a half years' survival with normal development. Arch Dis Childh 1984; 59:477–479.
479. Naughten ER, Flavin MP, O'Brien NG. A defect of the urea cycle-a case report. Irish J Med Sci 1984; 153:439–440.
480. Mayatepek E, Kurczynski TW, Hoppel CL, Gunning WT. Carnitine deficiency associated with ornthine transcarbamylase deficiency. Pediatr Neurol 1991; 7:196–199.
481. Maestri NE, McGowan KD, Brusilow SW. Plasma glutamine concentration: a guide in the management of urea cycle disorders. J Pediatr 1992; 121:259–261.
482. Bueno JD, Lutz R, Cho S. Ornithine transcarbamylase deficiency: case report and review. Kansas Med 1995; 96:135–138.
483. García-Pérez MA, Climent C, Briones P, Vilaseca MA, Rodés M, RubioV. Missense mutations in codon 225 of ornithine transcarbamylase (OTC) result in decreased amounts of OTC protein: a hypothesis on the molecular mechanism of the OTC deficiency. J Inher Metab Dis 1997; 20:769–777.
484. Busuttil AA, Goss JA, Seu P, Dulkanchainun TS, Yanni GS, McDiarmid SV, Busutttil RW. The role of orthotopic liver transplantation in the treatment of ornithine transcarbamylase deficiency. Liver Transpl Surg 1998; 4:350–354.
485. Brunquell P, Tezcan K, DiMario FJ Jr. Electroencephalographic findings in ornithine transcarbamylase deficiency. J Child Neurol 1999; 14:533–536.
486. Levin B, Dobbs RH. Hereditary metabolic disorders involving the urea cycle. Proc Roy Soc Med 1968; 61:773–774.
487. Levin B, Dobbs RH, Burgess EA, Palmer T. Hyperammonaemia. A variant type of deficiency of liver ornithine transcarbamylase. Arch Dis Childh 1969; 44:162–169.
488. MacLeod P, Mackenzie S, Scriver CR. Partial ornithine carbamyl transferase deficiency: an inborn error of the urea cycle presenting as orotic aciduria in a male infant. Canad Med Assoc J 1972; 107:405–408.
489. Aylsworth AS, Swisher CN, Kirkman HN. Lethal hyperammonemia due to partial ornithine transcarbamylase (OTC) deficiency in a six year old male. Pediatr Res 1975; 27:15A.
490. Van der Heiden C, Desplanque J, Bakker HD. Some kinetic properties of liver ornithine carbamoyl transferase (OCT) in a patient with OCT deficiency. Clin Chim Acta 1977; 80:519–527.
491. Qureshi IA, Letarte J, Ouellet R. Biochemical evaluation and dietary control of ornithine transcarbamylase (OTC) deficiency in a male child. Clin Res 1978; 26:865A.
492. Letarte J, Qureshi IA, Ouellet R, Godard M. Chronic benzoate therapy in a boy with partial ornithine transcarbamylase deficiency. J Pediatr 1985; 106:794–797.
493. Krieger I, Snodgrass PJ, Roskamp J. Atypical clinical course of ornithine

transcarbamylase deficiency due to a new mutant (comparison with Reye's disease). J Clin Endocrinol Metab 1979; 48:388–392.
494. Yudkoff M, Yang W, Snodgrass PJ, Segal S. Ornithine transcarbamylase deficiency in a boy with normal development. J Pediatr 1980; 96:441–443.
495. Schuchmann L, Colombo JP, Fischer H. Hyperammonämie durch ornithin-transcarbamylase-mangel-ursache letaler metabolischer krisen im säuglingsalter. Klin Pädiat 1980; 192:281–285.
496. Longhi R, Butte C, Valsasina R, Rossi L, Borzani M. Mitochondria abnormalities in a male with ornithine transcarbamylase deficiency (OTC). Pediatr Res 1981; 15:634A.
497. Yokoi T, Honke K, Funabashi T, Hayashi R, Suzuki Y, Taniguchi N, Hosoya M, Sakeki T. Partial ornithine transcarbamylase deficiency simulating Reye syndrome. J Pediatr 1981; 99:929–931.
498. Landrieu P, Baudouin F, Lyon G, VanHoof F. Liver peroxisome damage during acute hepatic failure in partial ornithine transcarbamylase deficiency. Pediatr Res 1982; 16:977–981.
499. Hoogenraad N, De Martinis L, Danks DM. Immunological evidence for an ornithine transcarbamylase lesion resulting in the formation of enzyme with smaller protein subunits. J Inher Metab Dis 1983; 6:149–152.
500. Tallan HH, Schaffner F, Taffet SL, Schneidman K, Gaull GE. Ornithine carbamoyl transferase deficiency in an adult male patient: significance of hepatic ultrastructure in clinical diagnosis. Pediatrics 1983; 71:224–232.
501. Oizumi J, Ng WG, Koch R, Shaw KNF, Sweetman L, Velazquez A, Donnell GN. Partial ornithine transcarbamylase deficiency associated with recurrent hyperammonemia, lethargy, and depressed sensorium. Clin Genet 1984; 25:538–542.
502. DiMagno EP, Lowe JE, Snodgrass PJ, Jones JD. Ornithine transcarbamylase deficiency—a cause of bizarre behavior in a man. N Engl J Med 1986; 315:744–747.
503. Stöckler S, Grosschadl F, Bachmann C, Roscher A. Ornithine transcarbamylase variant in a male patient. J Inher Metab Dis 1987; 10:272.
504. Drogari E, Leonard JV. Late onset ornithine carbamoyl transferase deficiency in males. Arch Dis Childh 1988; 63:1363–1367.
505. Coşkun T, Özalp I, Mönch S, Kneer J. Lethal hyperammonaemic coma due to ornithine transcarbamylase deficiency presenting as brain stem encephalitis in a previously asymptomatic ten-year-old boy. J Inher Metab Dis 1987; 10:271.
506. Snebold NG, Rizzo JF III, Lessell S, Pruett RC. Transient visual loss in ornithine transcarbamoylase deficiency. Am J Ophthalm 1987; 104:407–412.
507. Wendel U, Wieland J, Bremer HJ, Bachmann C. Ornithine transcarbamylase deficiency in a male:strict correlation between metabolic control and plasma arginine concentration. Eur J Pediatr 1989; 148:349–352.
508. Rabier D, Benoit A, Petit F, Chekoury A, Bonnefont JP, Saudubray JM,

Kamoun P. Ornithine carbamoyltransferase deficiency. A new variant with subnormal enzyme activity. Clin Chim Acta 1989; 186:25–30.
509. Kuno T, Miyazaki S, Inoue I, Saheki T. Hyperammonemia caused by impaired mitochondrial ornithine transport in a patient with partial quantitative deficiency of ornithine carbamoyltransferase. Clin Biochem 1990; 23:143–147.
510. Nishiyori A, Yoshino M, Tananari Y, Matsuura T, Hoshide R, Matsuda I, Mori M, Kato H. Y55D mutation in ornithine transcarbamylase associated with late-onset hyperammonemia in a male. Hum Mutat 1998; Suppl 1:S131–S133.
511. Finkelstein JE, Hauser ER, Leonard CO, Brusilow SW. Late-onset ornithine transcarbamylase deficiency in male patients. J Pediatr 1990a; 117:897–902.
512. Legius E, Baten E, Stul M, Marynen P, Cassiman J-J. Sporadic late onset ornithine transcarbamylase deficiency in a boy with somatic mosaicism for an intragenic deletion. Clin Genet 1990; 38:155–159.
513. Mizoguchi K, Sukehiro K, Ogata M, Onizuka S. Watanabe J, Yoshida I, Yoshino M. A case of ornithine transcarbamylase deficiency with acute and late onset simulating Reye's syndrome in an adult male. Kurume Med J 1990; 37:105–109.
514. Rabier D, Guillois B, Bardet J, Deprun C, Parvy P, Kamoun P. Ornithine carbamoyltransferase deficiency with subnormal enzyme activity. J Inher Metab Dis 1991b; 14:842–843.
515. Yoshida I, Yoshino M, Watanabe J, Yamashita F. Sudden onset of ornithine carbamoyltransferase deficiency after aspirin ingestion. J Inher Metab Des 1993; 16:917.
516. Brusilow SW, Finkelstein J. Restoration of nitrogen homeostasis in a man with ornithine transcarbamylase deficiency. Metabolism 1993; 42:1336–1339.
517. Wilson BE, Hobbs WN, Newmark JJ, Farrow SJ. Rapidly fatal hyperammonemic coma in adults. Urea cycle enzyme deficiency. West J Med 1994; 161:166–168.
518. Spada M, Guardamagna O, Rabier D, van der Meer SB, Parvy P, Bardet J, Ponzone A, Saudubray JM. Recurrent episodes of bizarre behavior in a boy with ornithine transcarbamylase deficiency: diagnostic failure of protein loading and allopurinol challenge tests. J Pediatr 1994; 125:249–251.
519. Capistrano-Estrada S, Marsden DL, Nyhan WL, Newbury RO,, Krous HF, Tuchman M. Histopathological findings in a male with late-onset ornithine transcarbamylase deficiency. Pediatr Pathol 1994; 14:235–243.
520. Myers JH, Shook JE. Vomiting, ataxia, and altered mental status in an adolescent: late-onset ornithine transcarbamylase deficiency. Am J Emerg Med 1996; 14:553–557.
521. Largilliére C, Farriaux JP. Ornithine transcarbamylase deficiency in a boy with long survival. J Pediatr 1988; 113:952.

522. Maestri NE, Clissold D, Brusilow SW. Neonatal onset ornithine transcarbamylase deficiency: a retrospective analysis. J Pediatr 1999; 134:268–272.
523. Harris ML, Oberholzer VG. Conditions affecting the colorimetry of orotic acid and orotidine in urine. Clin Chem 1980; 26:473–479.
524. Brusilow SW, Hauser E. Simple method of measurement of orotic acid and orotidine in urine. J Chromatog 1989; 493:388–391.
525. Nuzum CT, Snodgrass PJ. Multiple assays of the five urea cycle enzymes in human liver homogenates, The Urea Cycle. Edited by S. Grisolia, R. Baguena, F. Major. New York, John Wiley & Sons, 1976, pp. 325–349.
526. Tuchman M, Tsai MY, Holzknecht RA, Brusilow SW. Carbamyl phosphate synthetase and ornithine transcarbamylase activities in enzyme-deficient human liver measured by radiochromatography and correlated with outcome. Pediatr Res 1989; 26:77–82.
527. Cathelineau L, Saudubray JM, Polonovski C. Ornithine carbamyl transferase: the effects of pH on the kinetics of a mutant human enzyme. Clin Chim Acta 1972; 41:305–312.
528. Kamoun P, Rabier D. Choice of substrate concentration in the search for abnormal enzyme activity in human tissues. Clin Chem 1989; 35: 1267.
529. Nuzum CT, Snodgrass PJ. Urea cycle enzyme adaptation to dietary protein in primates. Science 172:1042–1043.
530. Matsuda I, Nagata N, Matsuura T, Oyanagi K, Narisawa K, Kitagawa T, Sakiyama T, Yamashita F, Yoshino M. Retrospective survey of urea cycle disorders: Part 1. Clinical and laboratory observations of thirty two Japanese male patients with ornithine transcarbamylase deficiency. Am J Med Genet 1991; 38:85–89.
531. Zhang W, Holzknecht RA, Butkowski RJ, Tuchman M. Immunochemical analysis of carbamyl phosphate synthetase 1 and ornithine transcarbamylase deficient livers: elevated N-acetylglutamate level in a liver lacking carbamyl phosphate synthetase protein. Clin Invest Med 1990; 4:183–188.
532. Bachmann C. Ornithine carbamoyl transferase deficiency: findings, models and problems. J Inher Metab Dis 1992; 15:578–591.
533. Cathelineau L, Saudubray J-M, Polonovski C. Heterogenous mutations of the structural gene of human ornithine carbamyltransferase as observed in five personal cases. Enzyme 1974; 18:103–113.
534. Cathelineau L, Briand P, Petit F, Nuyts J-P, Farriaux J-P, Kamoun PP. Kinetic analysis of a new human ornithine carbamoyltransferase variant. Biochim Biophys Acta 1980; 614:40–45.
535. Briand P, Francois B, Rabier D, Cathelineau L. Ornithine transcarbamylase deficiencies in human males. Kinetic and immunochemical classification. Biochim Biophys Acta 1982; 704:100–106.
536. Cavard C, Cathelineau L, Rabier D, Briand P. Immunochemical analysis of nineteen ornithine transcarbamoylase deficiencies. Enzyme 1988; 40:51–56.
537. Saheki T, Imamura Y, Inoue I, Miura S, Mori M, Ohtake A, Tatibana M,

Katsumata N, Ohno T. Molecular basis of ornithine transcarbamylase deficiency lacking enzyme protein. J Inher Metab Dis 1984; 7:2–8.
538. Tuchman M, Holzknecht RA. Heterogeneity of patients with late onset ornithine transcarbamylase deficiency. Clin Invest Med 1991; 14:320–324.
539. Brown GW Jr, Cohen PP. Comparative biochemistry of urea synthesis I. Methods for the quantitative assay of urea cycle enzymes in liver. J Biol Chem 1959; 234:1769–1774.
540. Raijman L. Ornithine transcarbamylase: measurement in liver. Methods of Enzymatic Analysis, Basel, Verlag Chemie Weinheim, 1981, pp. 326–334.
541. Yorifuji T, Muroi J, Uematsu A, Tanaka K, Kiwaki K, Endo F, Matsuda I, Nagasaka H, Furusho K. X-inactivation pattern in the liver of a manifesting female with ornithine transcarbamylase (OTC) deficiency. Clin Genet 1998; 54:349–353.
542. Kay JDS, Hilton-Jones D, Hyman N.Valproate toxicity and ornithine carbamoyltransferase deficiency. Lancet 1986; ii:1283–1284.
543. Ohtani Y, Ohyanagi K, Yamamoto S, Matsuda I. Secondary carnitine deficiency in hyperammonemic attacks of ornithine transcarbamylase deficiency. J Pediatr 1988; 112:409–414.
544. Honeycutt D, Callahan K, Rutledge L, Evans B. Heterozygote ornithine transcarbamylase deficiency presenting as symptomatic hyperammonemia during initiation of valproate therapy. Neurology 1992; 42:666–668.
545. Tripp JH, Hargreaves T, Anthony PP, Searle JF, Miller P, Leonard JV, Patrick AD, Oberholzer VG. Sodium valproate and ornithine carbamyl transferase deficiency. Lancet 1981; ii:1165–1166.
546. Kennedy CR, Cogswell JJ. Late onset ornithine carbamoyl transferase deficiency in males. Arch Dis Childh 1989; 64:638.
547. Coude FX, Rabier D, Cathelineau L, Grimber G, Parvy P, Kamoun PP. Letter to the editor: a mechanism for valproate-induced hyperammonemia. Pediatr Res 1981; 15:974–975.
548. Coude FX, Grimber G, Parvy P, Rabier D, Petit F. Inhibition of ureagenesis by valproate in rat hepatocytes. Role of N-acetylglutamate and acetyl-CoA. Biochem J 1983; 216:233–236.
549. Alonso E, Girbés J, García-España A, Rubio V. Changes in urea cycle-related metabolites in the mouse after combined administration of valproic acid and an amino acid load. Arch Biochem Biophys 1989; 272:267–273.
550. Lund P, Wiggins D. Inhibition of carbamoyl-phosphate synthase (ammonia) by Tris and Hepes. Effect on K_a for N-acetylglutamate. Biochem J 1987; 243:273–276.
551. Lund P, Wiggins D. Is N-acetylglutamate a short-term regulator of urea synthesis? Biochem J 1984; 218:991–994.
552. Martin-Requero A, Corkey BE, Cerdan S, Walajtys-Rode E, Parrilla RL, Williamson JR. Interactions between α-ketoisovalerate metabolism and the pathways of gluconeogenesis and urea synthesis in isolated hepatocytes. J Biol Chem 258; 3673–3681.
553. Gruskay JA, Rosenbeerg LE. Inhibition of hepatic mitochondrial carbamyl

phosphate synthetase (CPS 1) by acyl CoA esters: possible mechanism of hyperammonemia in the organic acidemias. Pediatr Res 1979; 13:475.
554. Qureshi IA, Letarte J, Tuchweber B, Yousef I, Qureshi SR. Hepatotoxicity of sodium valproate in ornithine transcarbamylase-deficient mice. Toxicol Lett 1985; 25:297–306.
555. Williams CA, Tiefenbach S, McReynolds JW. Valproic acid-induced hyperammonemia in mentally retarded adults. Neurology 1984; 34:550–553.
556. Patsalos PN, Wilson SJ, Popovic M, Cowan JMA, Shorvon SD, Hjelm M. The prevalence of valproic-acid-associated hyperammonaemia in patients with intractable epilepsy resident at the Chalfont Centre for Epilepsy. J Epilepsy 1993; 6:228–232.
557. Coulter DL, Allen RJ. Hyperammonemia witth valproic acid therapy. J Pediatr 1981; 99:317–319.
558. Altunbaşak S, BaytokV, Tasouji M, Hergüner O, Burgut R, Kayrin L. Asymptomatic hyperammonemia in children treated with valproic acid. J Child Neurol 1997; 12:461–463.
559. Raby WN. Carnitine for valproic acid-induced hyperammonemia. Am J Psychiatry 1997; 154:1168–1169.
560. Gidal BE, Inglese CM, Meyer JF, Pitterle ME, Antonopolous J, Rust RS. Diet- and valproate-induced transient hyperammonemia: effect of L-carnitine. Pediatr Neurol 1997; 16:301–305.
561. Willmore LJ, Wilder BJ, Bruni J, Villarreal HJ. Effect of valproic acid on hepatic function. Neurology 1978; 28:961–964.
562. Dreifuss FE, Santilli N, Langer DH, Sweeney KP, Moline KA, Menander KB. Valproic acid hepatic fatalities: a retrospective review. Neurology 1987; 37:379–385.
563. Suchy FJ, Balistreri WF, Buchino JJ, Sondheimer JM, Bates SR, Kearns GL, Stull JD, Bove KE. Acute hepatic failure associated with the use of sodium valproate. Report of two fatal cases. New Engl J Med 1979; 300:9962–966.
564. Batshaw ML, Brusilow SW. Valproate-induced hyperammonemia. Ann Neurol 1982; 11:319–321.
565. Triggs WJ, Bohan TP, Lin S-N, Willmore LJ. Valproate-induced coma with ketosis and carnitine insufficiency. Arch Neurol 1990; 47:1131–1133.
566. Rett A. Über ein zerebral-atrophisches syndrom bei hyperammonämie.1966, Brüder Hollinek, Vienna, pp. 7–68.
567. Hagberg B, Aicardi J, Dias K, Ramos O. A progressive syndrome of autism, dementia, ataxia, and loss of purposeful hand use in girls: Rett's syndrome: report of 35 cases. Ann Neurol 1983; 14:471–479.
568. Hagberg B, Goutières F, Hanefeld F, Rett A, Wilson J. Rett syndrome: criteria for inclusion and exclusion. Brain Dev 1985; 7:372–373.
569. Bachmann C, Colombo JP, Gugler E, Kilian W, Rett A, da Silva V. Biotin and Rett syndrome. Am J Med Genet 1986; 24;323–330.
570. Thomas S, Hjelm M, Oberholzer V, Brett EM, Wilson J. Rett's syndrome

and ornithine carbamoyltransferase deficiency. Lancet 1987; ii:1330–1331.
571. Carpenter KH, Bonham JR, Clarke A. Rett's syndrome and ornithine carbamoyltransferase deficiency. J Inher Metab Dis 1990; 13: 308–310.
572. Thomas S, Oberholzer V, Wilson J, Hjelm M. The urea cycle in the Rett syndrome. Brain Dev 1990; 12:93–96.
573. Cameron D, Losty H, Wallace S. The Rett syndrome and ornithylcarbamoyl transferase deficiency. Brain Dev 1991; 13:138.
574. Hyman SL, Batshaw ML. A case of ornithine transcarbamylase deficiency with Rett syndrome manifestations. Am J Med Genet 1986; 24:339–343.
575. Amir RE, Van den Veyver IB, Wan M, Tran CQ, Francke U, Zoghbi HY. Rett syndrome is caused by mutations in X-linked MECP2, encoding methyl-CpG-binding protein 2. Nature Genet 1999; 23:185–188.
576. Dunn HG, MacLeod PM. Rett syndrome: review of biological abnormalities. Can J Neurol Sci 2001; 28:16–29.

8. Diagnosis and Treatment of OTC Deficiency

577. Urea cycles disorders conference group. Consensus statement from a conference for the management of patients with urea cycle disorders. J Pediatr 2001; 138:S1–S5.
578. Summar M, Tuchman M. Proceedings of a consensus conference for the management of patients with urea cycle disorders. J Pediatr 2001; 138:S6–S10.
579. Huizenga JR, Tangerman A, Gips CH. Determination of ammonia in biological fluids. Ann Clin Biochem 1994: 31:529–543.
580. Huizenga JR, Tangerman A, Gips CH. A rapid method for blood ammonia determination using the new blood ammonia checker (BAC) II. Clin Chim Acta 1992; 210:153–155.
581. Barsotti RJ. Measurement of ammonia in blood. J Pediatr 2001; 138:S11–S20.
582. Hudak ML, Jones MD Jr, Brusilow SW. Differentiation of transient hyperammonemia of the newborn and urea cycle enzyme defects by clinical presentation. J Pediatr 1985; 107:712–719.
583. Hoffmann G,Aramaki S, Blum-Hoffmann E, Nyhan WL, Sweetman L. Quantitative analysis for organic acids in biological samples: batch isolation followed by gas chromatographic-mass spectrometric analysis. Clin Chem 1989; 35:587–595.
584. Hoffmann GF. Organic acid analysis. In: Blau N, Duran M, Blaskovics ME, editors. Physician's guide to the laboratory diagnosis of metabolic diseases. London, England: Chapman & Hall Medical; 1996. p. 42–48.
585. Shih VE. Amino acid analysis. Ibid., pp. 13–29.
586. Steiner RD, Cederbaum SD. Laboratory evaluation of urea cycle disorders. J Pediatr 2001; 138:S21–S29.

587. Bonham JR, Guthrie P, Downing M, Allen JC,Tanner MS, Sharrard M, Rittey C, Land JM, Fensom A, O'Neill D, Duley JA, Fairbanks LD. The allopurinol load test lacks specificity for primary urea cycle defects but may indicate unrecognized mitochondrial disease. J Inher Metab Dis 1999; 22:174–184.

588. Carpenter KH, Potter M, Hammond JW, Wilcken B. Benign persistent orotic aciduria and the possibility of misdiagnosis of ornithine carbamoyltransferase deficiency. J Inher Metab Dis 1997; 20:354–358.

589. Wood MH, O'Sullivan WJ. The orotic aciduria of pregnancy. Am J Obstet Gynecol 1973; 116:57–61.

590. Baumgartner MR, Hu CA, Almashanu S, Steel G, Obie C, Aral B, Rabier D, Kamoun P, Saudubray J-M, Valle D. Hyperammonemia with reduced ornithine, citrulline, arginine and proline: a new inborn error caused by a mutation in the gene encoding Δ^1-pyrroline-5-carboxylate synthase. Hum Molec Genet 2000; 9:2853–2858.

591. Snodgrass PJ. Biochemical aspects of urea cycle disorders. Pediatrics 1981; 68:273–283.

592. Boyde TRC, Rahmatullah M. Optimization of conditions for the colorimetric determination of citrulline, using diacetyl monoxime. Anal Biochem 1980; 107:424–431.

593. Matsushima A, Orii T. The activity of carbamoyl-phosphate synthetase 1 and ornithine carbamoyltransferase (OCT) deficiency in the rectal mucosa. J Inher Metab Dis 1981; 4:83–84.

594. Cathelineau L, Briand P, Rabier D, Navarro J. Ornithine transcarbamylase and disaccharidase activities in damaged intestinal mucosa of children-Diagnosis of hereditary ornithine transcarbamylase deficiency in mucosa. J Pediatr Gastroent Nutr 1985; 4:960–964.

595. Wolfe DM, Gatfield PD. Leukocyte urea cycle enzymes in hyperammonemia. Pediatr Res 1975; 9:531.

596. Snodgrass PJ, Wappner RS, Brandt IK. Letter to the editor: white cell ornithine transcarbamylase activity cannot detect the liver enzyme deficiency. Pediatr Res 1978; 12:873.

597. Nagata N, Akaboshi I, Yamamoto J, Matsuda I, Ohtsuka H, Katsuki T. Ornithine transcarbamylase (OTC) in white blood cells. Pediatr Res 1980; 14:1370–1373.

598. McLaren J, Ng WG. Assay of ornithine carbamoyltransferase activity in human liver using carbon-labeled ornithine and thin-layer chromatography. Clin Chim Acta 1977; 81:193–201.

599. Karsai T, Ménes A, Molnár J, Elödi P. Determination of enzyme activity by chromatography and videodensitometry. II. Urea cycle enzymes in tissue homogenates. Acta Biochim Biophys Acad Sci Hung 1979; 14:133–142.

600. Watanabe Y, Mori S, Ozaki M, Fujiyama S, Sato T, Mori M. A sensitive enzyme-linked immunosorbent assay of serum ornithine carbamoyltransferase. Enzyme Prot 1994–5; 48:10–17.

601. Viglio S, Valentini G, Zanaboni G, Cetta G, De Gregorio A, Iadarola P. Rapid detection of ornithine transcarbamylase activity by micellar electrokinetic chromatography. Electrophoresis 1999; 20:138–144.
602. Yudkoff M, Daikhin Y, Ye X, Wilson JM, Batshaw ML. *In vivo* measurement of ureagenesis with stable isotopes. J Inher Metab Dis 1998; 21(Suppl 1):21–29.
603. Yudkoff M, Daikhin Y, Nissim I, Jawad A, Wilson J, Batshaw M. In vivo nitrogen metabolism in ornithine transcarbamylase deficiency. J Clin Invest 1996; 98:2167–2173.
604. Lee B, Yu H, Jahoor F, O'Brien W, Beaudet AL, Reeds P. *In vivo* urea cycle flux distinguishes and correlates with phenotypic severity in disorders of the urea cycle. Proc Natl Acad Sci USA 2000; 97:8021–8026.
605. Brusilow SW, Hauser E. Simple method of measurement of orotic acid and orotidine in urine. J Chromatog 1989; 493:388–391.
606. Banditt P. Determination of orotic acid in serum by high-performance liquid chromatography. J Chromatog B 1994; 660:176–179.
607. Seiler N, Grauffel C, Therrien G, Sarhan S, Knoedgen B. Determination of orotic acid in urine. J Chromatog B 1994; 653:87–91.
608. Sebesta I, Fairbanks LD, Davies PM, Simmonds HA, Leonard JV. The allopurinol loading test for identification of carriers for ornithine carbamoyl transferase deficiency: studies in a healthy control population and females at risk. Clin Chim Acta 1994; 224:45–54.
609. Rimoldi M, Bergomi P, Romeo A, DiDonato S. A new stable-isotope dilution method for measurement of orotic acid utilizing solvent-extracted urine. J Inher Metab Dis 1994; 17:243–244.
610. McCann MT, Thompson MM, Gueron IC, Tuchman M. Quantification of orotic acid in dried filter-paper urine samples by stable isotope dilution. Clin Chem 1995; 41:739–743.
611. Qureshi IA, Letarte J, Lebel S, Ouellet R. Variabilité de l'activité enzymatique et de l'acidurie orotique chez les souris *spf/+* hétérozygotes déficientes en ornithine transcarbamylase. Diabete Metab 1986; 12:250–255.
612. Snodgrass PJ. Unpublished observations,1990.
613. Maestri NE, Lord C, Glynn M, Bale A, Brusilow SW. The phenotype of ostensibly healthy women who are carriers for ornithine transcarbamylase deficiency. Medicine 1998; 77:389–397.
614. Ng WG, Oizumi J, Koch R, Shaw KNF, McLaren J, Donnel GN, Carter M. Carrier detection of urea cycle disorders. Pediatrics 1981; 68:448–452.
615. Haan EA, Danks DM, Grimes A. Carrier detection in ornithine transcarbamylase deficiency. J Inher Metab Dis 1982; 5:37–40.
616. Becroft DMO, Barry DMJ, Webster DR, Simmonds HA. Failure of protein loading tests to identify heterozygosity for ornithine carbamoyltransferase deficiency. J Inher Metab Dis 1984; 7:157–159.
617. MacKenzie AE, MacLeod HL, Heick HMC, Korneluk RG. False positive

results from the alanine loading test for ornithine carbamoyltransferase deficiency heterozygosity. J Pediatr 1989; 115:605–608.

618. Potter M, Hammond JW, Sim K-G, Green AK, Wilcken B. Ornithine carbamoyltransferase deficiency: improved sensitivity of testing for protein tolerance in the diagnosis of heterozygotes. J Inher Metab Dis 2001; 24:5–14.
619. Burlina AB, Ferrari V, Dionisi-Vici C, Bordugo A, Zacchello F, Tuchman M. Allopurinol challenge test in children. J Inher Metab Dis 1992; 15:707–712.
620. Arranz JA, Riudor E, Rodés M, Roig M, Climent C, Rubio V, Sentis M, Burlina A. Optimization of allopurinol challenge: sample purification, protein intake control, and the use of orotidine response as a discriminative variable improve performance of the test for diagnosing ornithine carbamoyltransferase deficiency. Clin Chem 1999; 45:995–1001.
621. Olier J, Gallego J, Digon E. Computerized tomography in primary hyperammonemia. Neuroradiol 1989; 31:356–367.
622. Verma NP, Hart ZH, Kooi KA. Electroencephalographic findings in urea-cycle disorders. Electroencephalography Clin Neurophys 1984; 57:105–112.
623. Latham PS, LaBrecque DR, McReynolds JW, Klatskin G. Liver ultrastructure in mitochondrial urea cycle enzyme deficiencies and comparison with Reye's syndrome. Hepatol 1984; 4:404–407.
624. Badizadegan K, Perez-Atayde AR. Focal glycogenosis of the liver in disorders of ureagenesis: its occurrence and diagnostic significance. Hepatol 1997; 26:365–373.
625. Zimmer K-P, Matsuda I, Matsuura T, Mori M, Colombo J-P, Fahimi HD, Koch H-G, Ullrich K, Harms E. Ultrastructural, immunocytochemical and stereological investigation of hepatocytes in a patient with the mutation of the ornithine transcarbamylase gene. Eur J Cell Biol 1995; 67:73–83.
626. Kornfeld M, Woodfin BM, Papile L, Davis LE, Bernard LR. Neuropathology of ornithine carbamyl transferase deficiency. Acta Neuropathol (Berl)1985;65:261–264.
627. Filloux F, Townsend JJ, Leonard C. Ornithine transcarbamylase deficiency: neuropathologic changes acquired in utero. J Pediatr1986;108: 942–945.
628. Summar M. Current strategies for the management of neonatal urea cycle disorders. J Pediatr 2001; 138:S30–S39.
629. Braun MC, Welch TR. Continuous venovenous hemodiafiltration in the treatment of acute hyperammonemia. Am J Nephrol 1998; 18:531–533.
630. Brusilow SW, Valle DL, Batshaw ML. New pathways of nitrogen excretion in inborn errors of urea synthesis. Lancet 1979; ii:452–454.
631. Brusilow S, Tinker J, Batshaw ML. Amino acid acylation: a mechanism of nitrogen excretion in inborn errors of urea synthesis. Science 1980; 207:659–661.
632. Tremblay GC, Qureshi IA. The biochemistry and toxicology of benzoic

acid metabolism and its relationship to the elimination of waste nitrogen. Pharmac Ther 1993; 60:63–90.
633. Moldave K, Meister A. Synthesis of phenylacetylglutamine by human tissue. J Biol Chem 1957; 229:463–476.
634. Brusilow SW. Phenylacetylglutamine may replace urea as a vehicle for waste nitrogen excretion. Pediatr Res 1991; 29:147–150.
635. Brusilow SW. Arginine, an indispensable amino acid for patients with inborn errors of urea synthesis. J Clin Invest 1984; 74:2144–2148.
636. Berry GT, Steiner RD. Long-term management of patients with urea cycle disorders. J Pediatr 2001; 138:S56–S61.
637. Leonard JV. The nutritional management of urea cycle disorders. J Pediatr 2001; 138:S40–S45.
638. Matsuda I, Ohtani Y, Ohyanagi K, Yamamoto S. Hyperammonimia related to carnitine metabolism with particular emphasis on ornithine transcarbamylase deficiency. Enzyme 1987; 38:251–255.
639. Batshaw ML, MacArthur RB, Tuchman M. Alternative pathway therapy for urea cycle disorders: twenty years later. J Pediatr 2001; 138:S46–S55.
640. Burlina AB, Ogier H, Korall H, Trefz FK. Long-term treatment with sodium phenylbutyrate in ornithine transcarbamylase-deficient patients. Mol Genet Metab 2001; 72:351–355.
641. Todo S, Starzl TE, Tzakis A, Benkov KJ, Kalousek F, Saheki T, Tanikawa K, Fenton WA. Orthotopic liver transplantation for urea cycle enzyme deficiency. Hepatol 1992; 15:419–422.
642. Lee B, Goss J. Long-term correction of urea cycle disorders. J Pediatr 2001; 138:S62–S71.
643. Broelsch CE, Emond JC, Whitington PF, Thistlethwaite JR, Baker AL, Lichtor JL. Application of reduced-size liver transplants as split grafts, auxiliary orthotopic grafts, and living related segmental transplants. Ann Surg 1990; 212:368–375.
644. Jan D, Poggi F, Jouvet P, Rabier D, Laurent J, Beringer A, Hubert P, Saudubray JM, Revillon Y. Definitive cure of hyperammonemia by liver transplantation in urea cycle defects: report of three cases. Transplant Proc 1994; 26:188.
645. Whitington PF, Alonso EM, Boyle JT, Molleston JP, Rosenthal P, Emond JC, Millis JM. Liver transplantation for the treatment of urea cycle disorders. J Inher Metab Dis 1998; 21 (Suppl 1):112–118.
646. Kiuchi T, Edamoto Y, Kaibori M, Uryuhara K, Kasahara M, Uemoto S, Egawa H, Inomata Y, Tanaka K. Auxiliary liver transplantation for urea-cycle enzyme deficiencies: lessons from three cases. Transplant Proc 1999; 31:528–529.
647. Saudubray JM, Touati G, Delonlay P, Jouvet P, Narcy C, Laurent J, Rabier D, Kamoun P, Jan D, Revillon Y. Liver transplantation in urea cycle disorders. Eur J Pediatr 1999; 158 (Suppl 2):S55–S59.
648. Scaglia F, Zheng Q, O'Brien WE, Henry J, Rosenberger J, Reeds P, Lee B.

An integrated approach to the diagnosis and prospective management of partial ornithine transcarbamylase deficiency. Pediatrics 2002; 109:150–152.

649. Cederbaum JA, LeMons C, Rosen M, Ahrens M, Vonachen S, Cederbaum SD. Psychosocial issues and coping strategies in families affected by urea cycle disorders. J Pediatr 2001; 138:S72–S80.
650. Batshaw ML, Wilson JM, Raper S, Yudkkoff M, Robinson MB. Clinical protocol. Recombinant adenovirus gene transfer in adults with partial ornithine transcarbamylase deficiency (OTCD). Human Gene Ther 1999;10:2419–2437.
651. Barbour V. The balance of risk and benefit in gene-therapy trials. Lancet 2000; 355:384.
652. Jenks S. Gene therapy death- 'Everyone has to share in the guilt'. J Natl Cancer Inst 2000; 92: 98–100.
653. Bartholomew DW, McClellan JM. A novel missense mutation in the human ornithine transcarbamylase gene. Hum Mutat 1998; 12:220.
654. Mavinakere M, Morizono H, Shi D, Allewell NM, Tuchman M. The clinically variable R40H mutant ornithine carbamoyltransferase shows cytosolic degradation of the precursor protein in CHO cells. J Inherit Metab Dis 2001; 24:614–622.
655. Genet S, Cranston T, Middleton-Price HR. Mutation detection in 65 families with a possible diagnosis of ornithine carbamoyltransferase deficiency including 14 novel mutations. J Inherit Metab Dis 2000; 23:669–676.
656. Galloway PJ, MacPhee GB, Galea P, Robinson PH. Severe hyperammonaemia in a previously healthy teenager. Ann Clin Biochem 2000; 37:727–728.
657. Schultz REH, Salo MK. Under recognition of late onset ornithine transcarbamylase deficiency. Arch Dis Child 2000; 82:390–391.
658. Giorgi M, Morrone A, Donati MA, Ciani F, Bardelli T, Biasucci G, Zammarchi E. Lymphocyte mRNA analysis of the ornithine transcarbamylase gene in Italian OTCD male patients and manifesting carriers: identification of novel mutations. Hum Mutat 2000; 15:380–381.
659. Brusilow SW, Horwich AL. Urea cycle enzymes. In: Scriver CR, Beaudet AL, Sly WS, Valle D, Childs B, Kinzler KW, Vogelstein B editors. The metabolic & molecular bases of inherited disease. 8th edition. New York: McGraw-Hill; 2001. p. 1909–1963.
660. Climent C, Rubio V. Identification of seven novel missense mutations, two splice-site mutations, two microdeletions and a polymorphic amino acid substitution in the gene for ornithine transcarbamylase (OTC) in patients with OTC deficiency. Hum Mutat 2002; 19:185–186.
661. Tuchman M, Jaleel N, Morizono H, Sheehy L, Lynch MG. Mutations and polymorphisms in the human ornithine transcarbamylase gene. Hum Mutat 2002; 19:93–107.

9. Induction and Suppression of OTC and Urea Cycle Enzymes in Bacteria, Fungi and Mammals

662. Vogel HJ, Vogel RH. Enzymes of arginine biosynthesis and their repressive control. Adv Enzymol 1974; 40:65–90.
663. Maas WK. The regulation of arginine biosynthesis: its contribution to understanding the control of gene expression. Genetics 1991; 128:489–494.
664. Davis RH. Compartmental and regulatory mechanisms in the arginine pathways of *Neurospora crassa* and *Saccharomyces cerevisiae*. Microbiol Rev 1986; 50:280–313.
665. Tuchman M. Inherited hyperammonemia. In: Blau N, Duran M, Blaskovics ME, editors. Physician's guide to the laboratory diagnosis of metabolic diseases. London: Chapman & Hall Medical; 1996. pp. 209–222.
666. Tatibana M, Shigesada K. Two carbamyl phosphate synthetases of mammals: specific roles in control of pyrimidine and urea biosynthesis. Adv Enzyme Regul 1972; 10:249–271.
667. Jackson MJ, Allen SJ, Beaudet AL, O'Brien WE. Metabolite regulation of argininosuccinate synthetase in cultured human cells. J Biol Chem 1988; 263:16388–16394.
668. Bruhat A, Jousse C, Fafournoux P. Amino acid limitation regulates gene expression. Proc Nutr Soc 1999; 58:625–632.
669. Bruhat A, Jousse C, Carraro V, Reimold AM,, Ferrara M, Fafournoux P. Amino acids control mammalian gene transcription: activating transcription factor 2 is essential for the amino acid responsiveness of the *CHOP* promoter. Molec Cell Biol 2000; 20:7192–7204.
670. Schimke RT. The importance of both synthesis and degradation in the control of arginase levels in rat liver. J Biol Chem 1964; 239:3808–3817.
671. Schimke RT. Differential effects of fasting and protein-free diets on levels of urea cycle enzymes in rat liver. J Biol Chem 1962; 237:1921–1924.
672. Pitot HC, Peraino C. Studies on the induction and repression of enzymes in rat liver. I. Induction of threonine dehydrase and ornithine-δ-transaminase by oral intubation of casein hydrolysate. J Biol Chem 1964;239:1783–1788.
673. Ibid. Studies on the induction and repression of enzymes in rat liver. II. Carbohydrate repression of dietary and hormonal induction of threonine dehydrase and ornithine δ-transaminase. J Biol Chem 1964; 239:4308–4313.
674. Snodgrass PJ. Unpublished observations, 1978.
675. Sudilovsky O, Pitot HC. Studies on the role of adenosine 3′,5′-monophosphate during glucose repression in rat liver. Proc Soc Exp Biol Med 1973; 144:113–121.

676. Greengard O, Dewey HK. The effects of glucose ingestion on basal and induced enzyme levels in rat tissues. Biochem Biophys Acta 1973; 329:241–250.
677. Seitz HK. Müller MJ, Nordmeyer P, Krone W, Tarnowski W. Concentration of cyclic AMP in rat liver as a function of the insulin/glucagon ratio in blood under standardized physiological conditions. Endocrinol 1976; 99:1313–1318.
678. McLean P, Novello F. Influence of pancreatic hormones on enzymes concerned with urea synthesis in rat liver. Biochem J 1965; 94:410–422.
679. Snodgrass PJ, Lin RC, Müller WA, Aoki TT. Induction of urea cycle enzymes of rat liver by glucagon. J Biol Chem 1978; 253:2748–2753.
680. McLean P, Gurney MW. Effect of adrenalectomy and of growth hormone on enzymes concerned with urea synthesis in rat liver. Biochem J 1963; 87:96–104.
681. Sochor M, McLean P, Brown J, Greenbaum AL. Regulation of pathways of ornithine metabolism. Effects of thyroid hormone and diabetes on the activity of enzymes at the "ornithine crossroads" in rat liver. Enzyme 1981: 26:15–23.
682. Eisenstein AB, Strack I, Gallo-Torres H, Georgiadis A, Miller ON. Increased glucagon secretion in protein-fed rats: lack of relationship to plasma amino acids. Am J Physiol 1979; 236:E20–E27.
683. Eisenstein AB, Strack I. Amino acid stimulation of glucagon secretion by perifused islets of high-protein-fed rats. Diabetes 1978; 27:370–376.
684. Rocha DM, Faloona GR, Unger RH. Glucagon-stimulating activity of 20 amino acids in dogs. J Clin Invest 1972; 51:2346–2351.
685. Wise JK, Hendler R, Felig P. Influence of glucocorticoids on glucagon secretion and plasma amino acid concentrations in man. J Clin Invest 1973; 52:2774–2782.
686. Mori M, Miura A, Tatibana M, Cohen PP. Cell-free translation of carbamyl phosphate synthetase I and ornithine transcarbamylase messenger RNAs of rat liver. Effect of dietary protein and fasting on translatable mRNA levels. J Biol Chem 1981; 256:4127–4132.
687. Morris SM Jr, Moncman CL, Rand KD, Dizikes GJ, Cederbaum SD, O'Brien WE. Regulation of mRNA levels for five urea cycle enzymes in rat liver by diet, cyclic AMP, and glucocorticoids. Arch Biochem Biophys 1987; 256:343–353.
688. Lin RC, Snodgrass PJ. Primary culture of normal adult rat liver cells which maintain stable urea cycle enzymes. Biochem Biophys Res Commun 1975; 64:725–734.
689. Gebhardt R, Mecke D. Permissive effect of dexamethasone on glucagon induction of urea-cycle enzymes in perifused primary monolayer cultures of rat hepatocytes. Eur J Biochem 1979; 97:29–35.
690. Lin RC, Snodgrass PJ, Rabier D. Induction of urea cycle enzymes by glu-

cagon and dexamethasone in monolayer cultures of adult rat hepatocytes. J Biol Chem 1982;257:5061–5067.
691. Nebes VL, Morris SM Jr. Regulation of messenger ribonucleic acid levels for five urea cycle enzymes in cultured rat hepatocytes. Requirements for cyclic adenosine monophosphate, glucocorticoids, and ongoing protein synthesis. Molec Endocrinol 1988; 2:444–451.
692. Morris SM Jr, Kepka-Lenhart D. Hormonal induction of hepatic mitochondrial ornithine/citrulline transporter mRNA. Biochem Biophys Res Commun 2002; 294:749–752.
693. Ulbright C, Snodgrass PJ. Coordinate induction of the urea cycle enzymes by glucagon and dexamethasone is accomplished by three different mechanisms. Arch Biochem Biophys 1993; 301:1–7.
694. Matsuno F, Chowdhury S. Gotoh T, Iwase K, Matsuzaki H, Takatsuki K, Mori M, Takiguchi M. Induction of the C/EBPβ gene by dexamethasone and glucagon in primary-cultured rat hepatocytes. J Biochem 1996; 119:524–532.
695. Kimura T, Christoffels VM, Chowdhury S, Iwase K, Matsuzaki H, Mori M, Lamers WH, Darlington GJ, Takiguchi M. Hypoglycemia-associated hyperammonemia caused by impaired expression of ornithine cycle enzyme genes in C/EBPα knockout mice. J Biol Chem 1998; 273:27505–27510.
696. Kimura T, Chowdhury S, Tanaka T, Shimizu A, Iwase K, Oyadomari S, Gotoh T, Matsuzaki H, Mori M, Akira S, Takiguchi M. CCAAT/enhancer-binding protein β is required for activation of genes for ornithine cycle enzymes by glucocorticoids and glucagon in primary-cultured hepatocytes. FEBS Lett 2001;494:105–111.
697. Tomomura M, Tomomura A, Abu Musa DMA, Saheki T. Long-chain fatty acids suppress the induction of urea cycle enzyme genes by glucocorticoid action. FEBS Lett 1996; 399:310–312.
698. Takiguchi M, Mori M. Transcriptional regulation of genes for ornithine cycle enzymes. Biochem J 1995; 312:649–659.
699. Illnerova H. The development of arginase and ornithine-transcarbamylasse activities in the liver of rats. Physiol Bohemoslovaca 1966; 15:19–22.
700. Lamers WH, Mooren PG. Multihormonal control of enzyme clusters in rat liver ontogenesis. 1. Effects of adrenalectomy and gonadectomy. Mechanisms Ageing Develop 1981; 15:77–92.
701. Ibid. 2. Role of glucocorticosteroid and thyroid hormone and of glucagon and insulin.
702. Lamers WH, Mooren PG, Oosterhuis W, Lunstroo H, DeGraaf A, Charles R. The relation between the developmental timing of birth and developmental increases in urea cycle enzymes. In: Adv Exper Med Biol, vol. 153; Lowenthal A, Mori A, Marescau B, editors. Urea cycle diseases. New York: Plenum Presss; 1982. p. 229–240.

703. Ryall JC, Quantz MA, Shore GC. Rat liver and intestinal mucosa differ in the developmental pattern and hormonal regulation of carbamoyl-phosphate synthetase I and ornithine carbamoyl transferase gene expression. Eur J Biochem 1986; 156:453–458.
704. Morris SM Jr, Kepka DM, Sweeney WE Jr, Avner ED. Abundance of mRNAs encoding urea cycle enzymes in fetal and neonatal mouse liver. Arch Biochem Biophys 1989; 269:175–180.
705. Mukarram Ali Baig M, Habibullah CM, Swamy M, Ibrahim Hassan S, Taher-uz-Zaman, Qamar Ayesha, Geetha Devi B. Studies on urea cycle enzyme levels in the human fetal liver at different gestational ages. Pediatr Res 1992; 31;1143–145.
706. Bourgeois P, Harlin J-C, Renouf S, Goutal I, Fairand A, Husson A. Regulation of argininosuccinate synthetase mRNA level in rat foetal hepatocytes. Eur J Biochem 1997; 249:669–674.
707. Räihä NCR. Developmental changes of urea-cycle enzymes in mammalian liver. In: Grisolia S, Baguena R, Mayor F, editors. The urea cycle. New York: John Wiley & Sons; 1976. p. 261–272.

Additional References

708. Lee Y, Yoo SK, Lee JS, Kwon YM. Genomic structure of ornithine carbamoyltransferase gene from *Canavalia lineata*. Ann Clin Biochem 2000; 37:727–728.
709. de Jongh HHJ. The helix nucleation site and propensity of the synthetic mitochondrial presequence of ornithine carbamoyltransferase. Eur J Biochem 2000; 267:5796–5804.
710. Iwata K, Kajimura M, Sakamoto T. Functional ureogenesis in the gobiid fish, *Mugilogobius abei*. J Exper Biol 2000; 203:3703–3715.
711. Shi D, Morizono H, Yu X, Tong L, Allewell NM, Tuchman M. Human ornithine transcarbamylase: crystallographic insights into substrate recognition and conformational changes. Biochem J 2001; 354:501–509.
712. De Gregorio A, Risitano A, Capo C, Criniò C, Petruzzilli R, Desideri A. Evidence of carbamoylphosphate induced conformational changes upon binding to human ornithine carbamoyltransferase. Biochem Mol Biol Int 1999; 47:965–970.
713. Shi D, Gallegos R, DePonte J III, Morizono H, Yu X, Allewell NM, Malamy M, Tuchman M. Crystal structure of a transcarbamylase-like protein from the anaerobic bacterium *Bacteroides fragilis* at 2.0 Å resolution. J Mol Biol 2002; 320:899–908.
714. Allewell NM, Shi D, Morizono H, Tuchman M. Molecular recognition by ornithine and aspartate transcarbamylases. Acc Chem Res 1999; 32;885–894.
715. Bogdanovic MD, Kidd D, Briddon A, Duncan JS, Land JM. Late onset het-

erozygous ornithine transcarbamylase deficiency mimicking complex partial status epilepticus. J Neurol Neurosurg Psychiatry 2000; 69:813–815.

716. Rapp B, Häberle J, Linnebank M, Wermuth B, Marquardt T, Harms E, Koch HG. Genetic analysis of carbamoylphosphate synthetase I and ornithine transcarbamylase deficiency using fibroblasts. Eur J Pediatr 2001; 160:283–287.
717. Barshop BA, Nyhan WL, Climent C, Rubio V. Pitfalls in the detection of heterozygosity by allopurinol in a variant form of ornithine carbamoyltransferase deficiency. J Inherit Metab Dis 2001; 24:513–514.
718. Ploechl E, Ploechl W, Stoeckler-Ipsiroglu S, Pokorny H, Wermuth B. Late-onset ornithine transcarbamylase deficiency in two families with different mutations in the same codon. Clin Genet 2001; 59:111–114.
719. Perpoint T, Argaud L, Blanc Q, Robert D. Fatal hyperammonemic coma caused by ornithine transcarbamylase deficiency in a woman. Intensive Care Med 2001; 27:1962.
720. Legras A, Labarthe F, Maillot F, Garrigue M-A, Kouatchet A, Ogier de Baulny H. Late diagnosis of ornithine transcarbamylase defect in three related female patients: polymorphic presentations. Crit Care Med 2002; 30:241–244.
721. Shi D, Morizono H, Yu X, Tong L, Allewell NM, Tuchman M. Crystallization and preliminary X-ray crystallographic studies of wild-type human ornithine transcarbamylase and two naturally occurring mutants at position 277. Acta Cryst 2001; D57:719–721.
722. Tuchman M, McCullough BA, Yudkoff M. The molecular basis of ornithine transcarbamylase deficiency. Eur J Pediatr 2000; 159{Suppl 3}:S196–S198.
723. DelloRusso C, Scott JM, Hartigan-O'Connor D, Salvatori G, Barjot C, Robinson AS, Crawford RW, Brooks SV, Chamberlain JS. Functional correction of adult *mdx* mouse muscle using gutted adenoviral vectors expressing full-length dystrophin. Proc Natl Acad Sci USA 2002; 99:12979–12984.
724. Yant SR, Ehrhardt A, Giehm Mikkelsen J, Meuse L, Pham T, Kay MA. Transposition from a gutless adeno-transposon vector stabilizes transgene expression *in vivo*. Nature Biotech 2002; 20:999–1005.
725. Wiltshire EJ, Poplawski NK, Harbord MG, Harrison RJ, Fletcher JM. Ornithine carbamoyltransferase deficiency presenting with chorea in a female. J Inherit Metab Dis 2000; 23:843–844.
726. Morra DR, Nadkarni VM,, Bartoshesky LE, Finkelstein MS. RC-a case of ornithine transcarbamylase (OTC) deficiency. The most commonly genetically acquired urea cycle defect. Del Med J 2000; 72:349–354.
727. Trevedi M, Zafar S, Spalding MJ, Jonnalagadda S. Ornithine transcarbamylase deficiency unmasked because of gastrointestinal bleeding. J Clin Gastroenterol 2001; 32:340–343.

728. Yamanouchi H, Yokoo H, Yuhara Y, Maruyama K, Sasaki A, Hirato J, Nakazato Y. An autopsy case of ornithine transcarbamylase deficiency. Brain & Development 2002; 24:91–94.
729. Wilson CJ, Lee PJ, Leonard JV. Plasma glutamine and ammonia concentrations in ornithine carbamoyltransferase deficiency and citrullinaemia. J Inherit Metab Dis 2001; 24:691–695.
730. Vossler DG, Wilensky AJ, Cawthon DF, Abson Kraemer DL, Ojemann LM, Caylor LM, Morgan JD. Serum and CSF glutamine levels in valproate-related hyperammonemic encephalopathy. Epilepsia 2002; 43:154–159.
731. Hamer HM, Knake S, Schomburg U, Rosenow F. Valproate-induced hyperammonemic encephalopathy in the presence of topiramate. Neurol 2000; 54:230–232.
732. O'Neill M, Dubrey RW, Grocott –Mason RM. Valproate encephalopathy and hyperammonaemia. Postgrad Med J 2002; 78:316–317.
733. Sumi S, Matsuura T, Kidouchi K, Togari H, Kubota M, Kitou O, Mikan H, Ohura T, Matsuda I, Wada Y. Detection of ornithine transcarbamylase deficiency heterozygotes by measuring of urinary uracil. Int J Mol Med 2000; 6:177–180.
734. Picca S, Dionisi-Vici C, Abeni D, Pastore A, Rizzo C, Orzalesi M, Sabetta G, Rizzoni G, Bartuli A. Extracorporeal dialysis in neonatal hyperammonemia: modalities and prognostic indicators. Pediatr Nephrol 2001; 16:862–867.
735. Hiroma T, Nakamura T, Tamura M, Kaneko T, Komiyama A. Continuous venovenous hemodiafiltration in neonatal onset hyperammonemia. Amer J Perinatol 2002; 19:221–224.
736. Bachmann C. Mechanisms of hyperammonemia. Clin Chem Lab Med 2002; 40:653–662.
737. Brusilow SW. Hyperammonemic encephalopathy. Medicine 2002; 81;240–249.
738. Butterworth RF. Glutamate transporters in hyperammonemia. Neuurochem Internat 2002; 41:81–85.
739. Corbalán R, Hernández-Viadel M, Llansola M, Montoliu C, Felipo V. Chronic hyperammonemia alters protein phosphorylation and glutamate receptor-associated signal transduction in brain. Neurochem Internat 2002; 41:103–108.
740. Felipo V, Butterworth RF. Mitochondrial dysfunction in acute hyperammonemia. Neurochem Internat 2002; 40:487–491.
741. Desjardins P, Butterworth RF. The "peripheral-type" benzodiazepine (omega 3) receptor in hyperammonemic disorders. Neurochem Internat 2002; 41:109–114.
742. Rao KV, Norenberg MD. Cerebral energy metabolism in hepatic encephalopathy and hyperammonemia. Metab Brain Dis 2001; 16:67–78.
743. Riudor E, Arranz JA, Rodés M, Rubio V, Sentís M, Burlina AB. Influence

of dose and age on the response of the alllopurinol test for ornithine carbamoyltransferase deficiency in control infants. J Inherit Metab Dis 2000; 23:662–668.
744. (Analysis.) Gene-therapy death prompts broad civil lawsuit. Nature Biotech 2000; 18:1136.
745. Raper SE, Yudkoff M, Chirmule N, Gao G-P, Nunes F, Haskal ZJ, Furth EE, Propert KJ, Robinson MB, Magosin S, Simoes H, Speicher L, Hughes J, Tazelaar J, Wivel NA, Wilson JM, Batshaw ML. A pilot study of *in vivo* liver-directed gene transfer with an adenoviral vector in partial ornithine transcarbamylase deficiency. Human Gene Ther 2002; 13:163–175.
746. Morris SM, Jr. Regulation of enzymes of the urea cycle and arginine metabolism. Annu Rev Nutr 2002; 22:87–105.
747. Krebs H. Reminiscences and reflections. Oxford (UK): Clarendon Press; 1981. p. 47.
748. Inoue Y, Hayhurst GP, Inoue J, Mori M, Gonzalez FJ. Defective ureagenesis in mice carrying a liver-specific disruption of hepatocyte nuclear factor 4α (HNF4α). J Biol Chem 2002; 277:25257–25265.
749. Kersten S, Mandard S, Escher P, Gonzalez FJ, Tafuri S, Desvergne B, Wahli W. The peroxisome proliferator-activated receptor α regulates amino acid metabolism. FASEB J 2001; 15:1971–1978.

Index

Biographical Note

Philip James Snodgrass joined the faculty of the Department of Medicine of Indiana University School of Medicine in 1973 as Professor of Medicine and Chief of the Medical Service at the R. L. Roudebush Veterans Administration Medical Center. He retired from the V.A. in 1995 and became Professor emeritus in 1999. He graduated from Harvard College in 1949 with a B.A. *cum laude* in biochemistry and from Harvard Medical School in 1953 *cum laude.* He obtained clinical training in internal medicine at the Peter Bent Brigham Hospital in Boston as intern, junior and senior resident, and as chief medical resident, 1961–1963. His research training in biochemistry and biophysics took place for two years as a Fellow of the National Foundation in the Biophysics Research Laboratory of the Peter Bent Brigham Hospital with Professor Bert L. Vallee and for one year with Dr. Warren E. C. Wacker at the Brigham Hospital as a Russell Stearns Research fellow. Dr. Snodgrass served as chief of the gastroenterology division of the Department of Medicine at the Brigham Hospital from 1963 to 1973, as an Associate in Medicine from 1963 to 1969 and as Assistant Professor of Medicine at Harvard Medical School from 1969 to 1973. He was a Visiting Scientist in the Metabolic Research Laboratory, Oxford University, in 1982. He has been a member of the American Society of Biological Chemistry and Molecular Biology and the American Institute of Nutrition, a Fellow of the American College of Physicians, and a member of the American Gastroenterology Association, the American Association for the Study of Liver Diseases and the American Pancreatic Association. Dr. Snodgrass's research has centered on the induction and suppression of urea cycle enzymes in mammalian liver and urea cycle enzyme deficiencies in humans and mutant mice. He has published on metalloenzymes, mitochondrial function, clinical chemistry, cholesterol metabolism, human liver glutaminase, and salmon body composition and has written textbook chapters on pancreatic, gallbladder and bile duct diseases.